Slip dynamics at a bimaterial interface

I0503600

A thesis presented

by

Ranjith Kunnath

to

The Division of Engineering and Applied Sciences

in partial fulfillment of the requirements

for the degree of

Doctor of Philosophy

in the subject of

Engineering Sciences

Harvard University

Cambridge, Massachusetts

May 2001

Thesis advisor   Author
James R. Rice   **Ranjith Kunnath**

Slip dynamics at a bimaterial interface

Abstract

The mathematical well-posedness of the problem of dynamic stability to perturbations from a state of steady frictional sliding along a planar interface between solids with
dissimilar elastic properties is studied. It has been recently discovered that this problem
is often ill-posed when a Coulomb friction law is taken to act on the interface, i.e. when
the frictional strength τ is proportional to the normal stress at the interface σ through the
(constant) friction coefficient f, $\tau = f\sigma$.

Two experimentally motivated features of friction that make the stability problem
well posed are identified. We show that a friction law of the form $\dot{\tau} = -(V/L)(\tau - f\sigma)$,
where V is the sliding velocity and L is a sliding length, regularizes the problem. The
distinguishing feature of this law is that it has a memory effect, and no instantaneous
effect, of normal stress on frictional strength. Such a response to normal stress changes is
suggested by high speed sliding experiments (with speeds of the order of 1 m/sec) of hard
steels against cutting tool materials. We also show that a friction law with an instantaneous
dependence of friction on sliding velocity in the form $\tau \sim a\sigma \ln(V)$ with $a > 0$ makes the
stability problem well-posed at low sliding speeds. Creep slippage experiments on a variety
of materials - metals, rocks, paper, polymers - at slip speeds of upto a few mm/sec support
such a positive, logarithmic instantaneous effect of sliding speed, as do models of thermally
activated creep at the contacting asperities of the two solids.

Slip stability in a simple spring block model is also studied. Analytical non-
linear stability results are derived for quasi-static slip of the block with an experimentally
motivated friction law dependent on the slip velocity as well as on the state of the sliding
surface. It is shown that the critical spring stiffness for stable steady sliding of the block
obtained from a linear analysis fully determines the non-linear stability results. We argue
that this implies a slip patch of a critical size has to develop in a continuum system before
an instability can develop.

Contents

Citations to Previously Published Work

Chapter 2 has been published in its entirety as

> Ranjith, K. and Rice, J. R., 1999. Stability of quasi-static slip in a single degree of freedom elastic system with rate and state dependent friction. J. Mech. Phys. Solids 47, 1207-1218.

All of Chapter 3 and parts of Chapter 4 are used in the publication

> Ranjith, K. and Rice, J. R., 2001. Slip dynamics at an interface between dissimilar materials. J. Mech. Phys. Solids 49, 341-361.

Parts of Chapter 4 also appear in

> Rice, J. R., Lapusta, N. and Ranjith, K., 2001. Rate and state dependent friction and the stability of sliding between elastically deformable bodies. J. Mech. Phys. Solids, In Press.

Acknowledgements

It is a pleasure to thank Professor Daniel Fisher for his contributions to this thesis. I have gained a lot by way of brief, yet illuminating discussions with Daniel, and by attending his lectures. I have also benefited enormously from interactions with the other members on my Ph.D. committee - Professors George Adams, John Hutchinson, L. Mahadevan, Jim Rice and Jeroen Tromp. I am thankful to them for their time and patience.

Dedicated to my parents

Chapter 1

Introduction

In this thesis, I present analytical studies of instabilities in the frictional sliding of solids in elastic surroundings. The interplay between elasticity, dynamics and friction is shown to produce a richness of phenomena both in continuum and in low-dimensional models.

The thesis has two independent parts:

- Slip stability in a spring block model (Chapter 2)

- Slip stability at an interface between dissimilar elastic solids (Chapters 3 and 4).

The focus of the thesis is on the latter part. Chapter 3 establishes the requisite framework and the main results are presented in Chapter 4. The current chapter provides an introduction and summarizes the results.

1.1 Introduction

Interfaces between dissimilar materials are ubiquitous in nature and in technology. In the earth, for example, material properties across a seismic fault often vary modestly due to plate motions over geological times. Sedimentary layers also often have striking differences in properties. In the technological world, composite materials comprising metal/polymer interfaces find wide application. Typically, the interface is the weakest part in these material systems and failure often happens by rupture of the interface. This motivates our study of rupturing along bimaterial interfaces. We limit ourselves to ruptures that are pure slip events, i.e., there is no opening at the interface. Crucial to understanding the dynamics of slip ruptures is a proper characterization of the frictional properties of the interface. An obvious point of interest is the velocity dependence of friction. In addition, slip at an interface between dissimilar materials alters the normal stress on the interface and therefore it is important to understand the effect of normal stress changes on friction also.

1.2 Background on friction

The most general dependence of frictional strength τ on normal stress σ and sliding velocity V can be written in the form

$$\tau = F(\sigma, V, \{\sigma\}, \{V\}) \tag{1.1}$$

where the curly brackets denote memory dependence.

The dependence on sliding speed at low speeds of up to a few mm/sec and at constant normal stress is well studied for a wide range of materials [Dieterich (1979, 1981), Ruina (1983), Tullis and Weeks (1986), Marone (1998)]. A logarithmic dependence of friction on sliding speed is seen that is generally attributed to thermally activated creep at asperity contacts. Furthermore, these experiments show a positive instantaneous dependence on velocity in this range of sliding velocities, i.e., $\partial F/\partial V > 0$. The dependence of friction, at constant normal stress, on the sliding velocity is often written empirically in the form

$$\tau = \tau_* + a\sigma \ln(V/V_*) + b\sigma \ln(V_*\theta/L) \tag{1.2}$$

where τ_* and V_* are reference values and θ is a state variable that describes the evolution of the sliding surface. a, b and L are material parameters and from the afore mentioned positivity of the instantaneous dependence on sliding velocity it follows that $a > 0$. a and

b are generally of order 0.02 to 0.04 for a wide range of materials and $|a - b|$ of order 0.01. Experiments by Linker and Dieterich (1992) and Richardson and Marone (1999) at slip rates of the order 1 μm/sec on the effect of varying normal stress show that in response to a step change in normal stress, $\Delta\sigma$ a partial strength change $(f - \alpha)\Delta\sigma$ is observed at the time of the step and a further memory-like change occurs with continuing slip over a few μm distance leading an ultimate change $f\Delta\sigma$.

While the precise form of the friction law incorporating both normal stress and velocity changes at these low speeds (up to a few mm/sec) is not well understood, a linearized friction law for perturbations from a reference state with slip velocity V_o, shear stress τ_o and normal stress σ_o can be written in the form

$$\tau - \tau_o = \frac{a\sigma_o}{V_o}(V - V_o) - \frac{b\sigma_o}{V_o}\int_0^t h_V(t')\dot{V}(t - t')dt' \quad (1.3)$$
$$+ (f - \alpha)(\sigma - \sigma_o) + \alpha\int_0^t h_\sigma(t')\dot{\sigma}(t - t')dt'$$

where $h_V(0) = h_\sigma(0) = 0$ and $h_V(t)$ and $h_\sigma(t) \to 1$ as $t \to \infty$, to incorporate the observed features. The first term on the right hand side represents the instantaneous dependence on sliding velocity perturbations. The second term represents the memory of prior history of sliding mediated by the function $h_V(t)$. Thus, a step change in sliding speed ΔV at constant normal stress σ_o causes an immediate change $a\sigma_o\Delta V/V_o$ in the frictional strength and an ultimate change of $(a - b)\Delta V/V_o$. Similar interpretations hold for the third and fourth terms above involving the normal stress dependence.

Earthquake ruptures, however, involve sliding speeds three orders higher than those studied in the above experiments. The form of the velocity dependence at such high sliding speeds is not currently well understood. Some experiments involving high sliding speeds of the order 1 to 10 m/sec have recently been carried out [Prakash and Clifton (1993), Frutschy and Clifton (1997), Prakash (1998)]. We make a preliminary attempt towards the description of the velocity dependence of friction at such high speeds in this thesis. We also focus on the normal stress dependence of friction at high speeds as studied in the experiments of Prakash and Clifton (1993) and Prakash (1998). In contrast to the low speed experiments, they show that at high speeds there is no instantaneous change of shear strength, but rather a gradual change which occurs over a few microns of sliding.

1.3   Background on elasticity

Here, a brief background on elasticity in the context of frictional stability problems is provided.

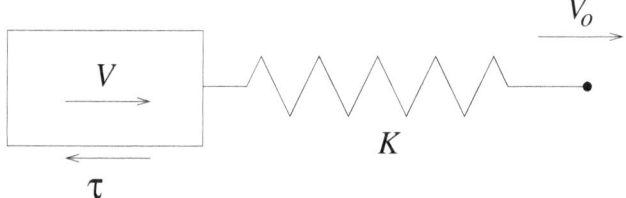

Figure 1.1: A spring block system sliding with imposed load-point motion

Consider first the simplest elastic system – a spring block system shown in Fig 1.1 with the block considered massless. A constant velocity V_o is impressed on the spring which has a stiffness K per unit area of sliding contact. Let V denote the slip velocity of the block, δ its slip displacement and τ the frictional shear stress acting on it. At steady state, the block slips with the same velocity V_o as the imposed motion and the frictional stress on the block is τ_o. When the slip is perturbed from its steady state value by an amount D, according to

$$\delta = V_o t + D, \tag{1.4}$$

equilibrium requires a shear stress perturbation given by

$$\tau - \tau_o = -KD. \tag{1.5}$$

Next, consider the quasi-static sliding of two elastic half-spaces past each other under anti-plane strain as shown in Fig 1.2. The interface is located at $x_2 = 0$ and the x_1 coordinate lies along the interface while slip and shear loading occur in the x_3 direction perpendicular to the plane of the figure. Let $u(x_1, x_2, t)$ denote the displacement field. As before, the slip velocity at steady state is V_o and the shear stress at the interface is τ_o. The equation of motion for quasi-static anti-plane motion is

$$\nabla^2 u = 0 \tag{1.6}$$

Consider the displacement field of the form

$$u(x_1, x_2, t) = \frac{1}{2} V_o t \ \mathrm{sign}(x_2) + \frac{1}{2} D(k, t) e^{ikx_1} e^{-|k||x_2|} \ \mathrm{sign}(x_2) \tag{1.7}$$

The first term on the right hand side represents the displacement field due to steady sliding at velocity V_o. The second term satisfies the Laplace equation and hence is an admissible displacement perturbation. The interfacial slip due to the above displacement field is

$$\delta = V_o t + D(k, t) e^{ikx_1} \tag{1.8}$$

4

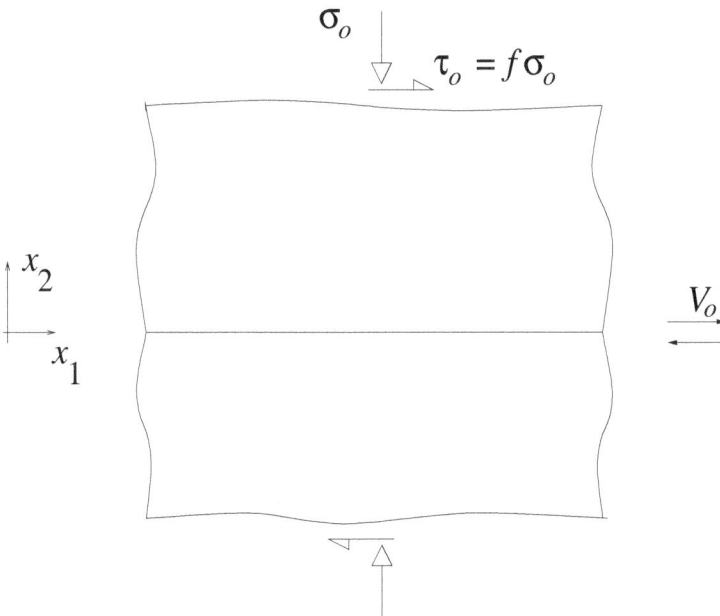

Figure 1.2: Anti-plane sliding along an interface between identical elastic solids

Using the elastic relation $\tau = \mu \partial u / \partial x_2$ for anti-plane strain, where μ is the shear modulus, the perturbation in the shear stress at the interface is given by

$$\tau - \tau_o = -\frac{1}{2}\mu \left|k\right| D(k,t)e^{ikx_1}. \qquad (1.9)$$

Comparing (1.4) with (1.8) and (1.5) with (1.9), we note that $\mu \left|k\right| /2$ acts like an effective stiffness of a continuum in anti-plane strain under quasi-static conditions. For in-plane sliding of solids with identical elastic properties, the corresponding quantity can be shown be $\mu \left|k\right| /2(1 - \nu)$, where ν is the Poisson's ratio. Thus short wavelengths are stiffer than long ones.

Consider next perturbations from a steady state of dynamic anti-plane sliding. The equation of motion for the displacement $u(x_1, x_2, t)$ then becomes

$$c_s^2 \nabla^2 u = \partial^2 u / \partial t^2, \qquad (1.10)$$

5

where c_s is the shear wave speed. For a slip perturbation of the form

$$\delta = V_o t + D(k,p)e^{ikx_1 + pt} \tag{1.11}$$

at the interface, the shear stress perturbation can be shown to be

$$\tau - \tau_o = -\frac{1}{2}\mu |k| (1 + p^2/k^2 c_s^2)^{1/2} D(k,p)e^{ikx_1 + pt} \tag{1.12}$$

following a similar analysis to the one presented above for quasi-static motions. We note immediately that a simple interpretation of the wavenumber as an effective stiffness is lost in the presence of dynamic effects. This sets the stage for our derivation in Chapter 3 of analogous relations for perturbations in shear and normal stress at a bimaterial interface due to imposed slip perturbations.

1.4 Summary of the thesis

The results of the individual chapters are summarized below.

In Chapter 2, the stability of quasi-static slip in a spring block model is studied, with slip following a rate and state dependent friction law. Alteration of normal stress is not considered. Linear stability analysis by Rice and Ruina (1983) identified the critical spring stiffness for stable motion of the block to be $K_{cr} = (b-a)\sigma/L$, where L is a characteristic length in the friction law. Our fully non-linear stability analysis shows that this parameter fully determines the structure of the non-linear stability results. For a constant imposed load point motion of the block, slip is stable when $K > K_{cr}$, irrespective of the strength of the perturbation from the steady state. When $K < K_{cr}$, slip motion is always unstable. When the load-point is held stationary and the block is perturbed by an initial condition, the value of K in relation to K_{cr} again determines whether accelerating creep motions are possible. If $K \geq K_{cr}$, such motions are ruled out whereas if $K < K_{cr}$, accelerating motions can occur in response to sufficiently large perturbations. For a fault patch of size D in elastic surroundings of stiffness μ, an effective stiffness can be identified as $K \sim \mu/D$. Thus, the existence of a critical stiffness for instability, irrespective of the size of a perturbation has the implication that the fault patch has to slip coherently to a critical size D_{cr} before it can break out unstably.

In Chapter 3, dynamic stability to perturbations from steady sliding at an interface between dissimilar elastic solids is studied. The key effect of material dissimilarity is that spatially non-uniform slip on the interface causes changes in normal stress acting on the interface. This has recently been recognized to give rise to unexpected dynamic instabilities.

In particular, it has been shown that the problem of stability to perturbations with a Coulomb law $\tau = f\sigma$, is often mathematically ill-posed due to a short wavelength instability. We show explicitly that the ill-posedness occurs at any arbitrarily small friction coefficient if the two solids constituting the bimaterial pair are such that an interfacial wave called the generalized Rayleigh wave exists for that material pair. This wave carries slip perturbations at a planar interface with no associated shear stress perturbations. In the limit of the materials becoming identical, its speed becomes precisely that of the Rayleigh wave. Hence it is called the generalized Rayleigh wave. It exists for modestly mismatched material pairs - for shear wave speed mismatches of up to 40% when the densities and Poisson's ratios of the two materials are not very different. We also show that intersonic unstable modes also exist quite often for reasonable values of the friction coefficient. When the generalized Rayleigh wave does not exist, we show that ill-posedness occurs for sufficiently high friction coefficients, $f > f_c$ and the unstable modes are always intersonic in that case. Our numerical results indicate that f_c can be surprisingly low for material systems of practical interest. For metal/polymer composite pairs, for example, even though the generalized Rayleigh does not exist, intersonic unstable modes mode exist when the friction coefficient f is more than about 0.03.

The main results of the thesis are in Chapter 4. We identify two corrections to the Coulomb friction law that regularize the problem of dynamic stability to perturbations from a steady sliding state for any given bimaterial pair. One regularizing law is of the form

$$\dot{\tau} = -(V/L)(\tau - f\sigma), \tag{1.13}$$

motivated by the experiments of Prakash and Clifton (1993) and Prakash (1998). According to this law, frictional strength has no direct dependence on normal stress, but only a memory dependence. Again, the existence of the generalized Rayleigh wave affects the stability results. If the wave does not exist, we are assured of absolute stability at short wave lengths (i.e., sufficiently short wavelength perturbations decay in time). However, if the generalized Rayleigh wave exists, the shortest wavelengths are only neutrally stable. We also show that a velocity dependent friction law of the form

$$\tau \sim a\sigma \ln(V) \tag{1.14}$$

makes the stability problem well-posed when slip speeds are low enough so that $\mu V/a\sigma c_s \ll 1$, where μ is a representative shear modulus and c_s a representative shear wave speed. Under this condition, it is shown that perturbations with sufficiently high wave number always die out. We have also obtained general results on the existence of interfacial waves and ordering of their speeds. We show that a generalized Rayleigh wave always exists when a Stonely wave exists and that the speed of the Stonely wave, when it exist, is always higher than that of the generalized Rayleigh wave.

1.5   Suggestions for future work

We have made an initial step in this thesis towards studying the problem of dynamic sliding along dissimilar material interfaces. The present work has identified two features of friction that make mathematically well-posed problems of stability to perturbations from steady sliding possible.

A larger problem of interest is understanding the modes of rupture propagation at bimaterial interfaces and the its dependence on friction and background stressing. Extensive work on this topic is reported in Lapusta (2001) for anti-plane sliding of solids with identical elastic properties. Two competing modes of rupture are the so-called self-healing pulse mode of rupture, where slip occurs for a small fraction of the duration of the rupture time, and the crack mode of rupture, where slip accumulates for times comparable to the duration of the rupture. A first step towards tackling this problem would be to do a fuller analysis to identify all instabilities and, especially, the ones with fastest growing rates (recall that we have focussed only on the short wavelength limit of the equations that govern stability). A further step that promises to be challenging mathematically is to study the effects of dispersion and of non-linearities in the friction on the dynamics. Such a study should give substantial predictive power to theory on the modes and structure of slip ruptures at bimaterial interfaces.

Another problem of interest is in identifying the microscopic origins of the velocity and normal stress dependence of friction and their roles in regularizing frictional stability problems. The chosen forms of the low speed laws have some theoretical backing from rate activation models at the contact asperities. However, a more fundamental understanding of the micromechanics of multiple asperity contacts is required to describe friction in terms of the asperity strengths and lifetimes. A starting point would be to study the mechanics of single asperity contacts and then to embody that in statistical description of multi-contact surfaces. There is very limited micromechanical understanding of frictional properties at high sliding speeds and it is not clear that the asperity contact models used to describe the low speed experiments would be valid. For example, the type of normal stress dependence seen in the high speed experiments of Prakash and Clifton (1993) and Prakash (1998) is not expected from asperity contact models, unless adhesion at contact is far greater at those

8

speeds, perhaps due to flash heating effects at contacts. When normal stress is changed suddenly, a corresponding instantaneous change in the asperity population is otherwise expected, giving rise to an instantaneous change in friction. However, such a change is not observed in the experiments. Further theoretical and experimental work to clarify these issues is merited.

Chapter 2

Stability of quasi-static slip in a single degree of freedom elastic system with rate and state dependent friction

The stability of quasi-static frictional slip of a single degree of freedom elastic system is studied for a Dieterich-Ruina rate and state dependent friction law, showing steady-state velocity weakening, and following the ageing (or slowness) version of the state evolution law. Previous studies have been done for the slip version.

Analytically determined phase plane trajectories and Liapunov function methods are used in this work. The stability results have an extremely simple form: (1) When a constant velocity is imposed at the load point, slip motion is always periodic when the elastic stiffness, K, has a critical value, K_{cr}. Slip is always stable when $K > K_{cr} > 0$, with rate approaching the load-point velocity, and unstable (slip rates within the quasi-static model become unbounded) when $K < K_{cr}$. This is unlike results based on the slip version of the state evolution law, in which instability occurs in response to sufficiently large perturbations from steady sliding when $K > K_{cr}$. An implication of this result for slip instabilities in continuum systems is that a critical nucleation size of coherent slip has to be attained before unstable slip can ensue. (2) When the load point is stationary, the system stably evolves towards slip at a monotonically decreasing rate whenever $K \geq K_{cr} > 0$. However, when $K < K_{cr}$, initial conditions leading to stable and unstable slip motion exist. Hence self-driven creep modes of instability exist, but only in the latter case.

2.1 Introduction

Consider a rigid block attached to a linear spring (Fig. 1.1). The block slides frictionally with velocity V when a constant velocity V_o is imposed at the other end of the spring. Normal stress on the contact interface is assumed to be constant. At steady state, the block slips with the velocity of the imposed motion. Away from steady state, the equation of motion of the block, assuming quasi-static slippage, is

$$\dot{\tau} = K(V_o - V) \tag{2.1}$$

where τ is the frictional shear stress on the block and K is the spring stiffness per unit area of sliding contact.

Motivated by experimental studies of rock friction, Dieterich (1979) and Ruina (1983) proposed an empirical frictional constitutive law in which friction depends both on the slip rate, V, and the state of the surface in the form

$$\tau = \tau_* + A \ln(V/V_*) + B \ln(V_* \theta / L) \tag{2.2}$$

where A, B and L are constants, τ_* and V_* are reference values of friction stress and sliding velocity, respectively, and θ is a state variable. Generally, τ_*, A and B are considered to be proportional to effective normal stress. Also, for V_* chosen in the range of imposed slip rates in the experiments mentioned (e.g., 10^{-9} to 10^{-3} m/s), A and B are typically of order 2% to 4% of τ_*, and $| A - B |$ of order 1% or less.

Based on the work of Dieterich (1979), Ruina (1983) introduced two widely used empirical laws for the evolution of the state variable. In one form, called the Dieterich-Ruina ageing (or slowness) law, the state variable is interpreted as an effective time of contact of surface asperities. State evolution by this law is described by the equation

$$\dot{\theta} = 1 - V\theta / L. \tag{2.3}$$

An important feature of this law is that friction evolves logarithmically with time even under stationary contact ($V = 0$). Hence it is called an ageing law. Another widely used state evolution law, referred to as the Ruina-Dieterich slip law is given by

$$\dot{\theta} = -(V\theta / L) \ln (V\theta / L) \tag{2.4}$$

Here, state evolves only during slip ($V \neq 0$). See Beeler et al. (1994), Perrin et al. (1995) and Rice and Ben-Zion (1996) for further discussions and comparisons of slip predictions based on these laws. The latter also discuss a physically based regularization (based on

Arrhenius rate process model) of the $\ln(V)$ term near $V = 0$ which is sometimes required, although not in the cases discussed here.

For both the Dieterich-Ruina ageing (or slowness) law and the Ruina-Dieterich slip law, at steady state,

$$V^{ss} = V_o, \quad \tau^{ss} = \tau_* - (B - A)\ln(V_o/V_*). \tag{2.5}$$

It is clear that the frictional stress at steady state decreases with increasing velocity when $B > A$.

A linear stability analysis of the steady state solution for both evolution laws was done by Ruina (1983), and more generally, for linear frictional constitutive laws with instantaneous velocity dependence and and fading memory of prior history of velocity, by Rice and Ruina (1983). They show that quasi-static steady state slip is stable ($V \to V_o$) or unstable ($V \to \infty$) as the spring stiffness is greater than or less than a critical value given by

$$K_{cr} = (B - A)/L \tag{2.6}$$

When $K = K_{cr}$, the linearized slip motion is periodic.

Gu et al. (1984) have done a non-linear analysis of the stability of steady slip with state evolution according to the Ruina-Dieterich slip law (2.4). For this law, they show that when $K = K_{cr}$, initial conditions leading both to periodic stick-slip motions as well as unstable motions exist. When $K > K_{cr}$, initial conditions sufficiently close to the steady state lead to stable slip while others cause unstable slip. When $K < K_{cr}$, slip is always unstable.

In the present work, a non-linear stability analysis of steady quasi-static slip, with evolution described by the Dieterich-Ruina ageing law (2.3), is carried out. The analysis, similar to the work of Gu et al. (1984), is carried out using analytically determined phase plane trajectories and Liapunov function techniques. The implications of the stability results for slip in continuum systems is also discussed.

This chapter is organized as follows: In Section 2.2, the governing equations are non-dimensionalized and a phase plane analysis is carried out. Analytical phase plane trajectories are derived for two particular cases: (1) the case when $K = K_{cr}$ with arbitrary, non-zero (constant) load point velocity, and (2) the case of a stationary load point with arbitrary spring stiffness. In Section 2.3, the stability of steady state slip with a non-stationary load point is studied using the phase plane trajectories obtained in Section 2.2 and by constructing a Liapunov function. Stability of sliding with a stationary load point

12

is studied in Section 2.4. The conditions under which unstable self-driven creep modes can exist are established and compared with those for the Ruina-Dieterich slip law. The results are finally summarized in Section 2.5.

2.2 Phase plane trajectories

The governing equations of the problem are (2.1), (2.2) and (2.3). We introduce the dimensionless quantities:

$$T = V_*t/L, \qquad k = KL/A, \quad \psi = (\tau - \tau_*)/A,$$
$$\phi = \ln(V/V_*), \quad \beta = B/A, \qquad v_o = V_o/V_*. \tag{2.7}$$

Combining (2.1), (2.2) and (2.3) to eliminate the state variable θ and using (2.7), we get

$$\frac{d\phi}{dT} = k(v_0 - e^\phi) - \beta \left[e^{(\phi-\psi)/\beta} - e^\phi \right] \tag{2.8}$$

$$\frac{d\psi}{dT} = k(v_o - e^\phi) \tag{2.9}$$

Now, (2.8) and (2.9) are the governing equations in non-dimensional form.

The steady state solution (2.5) can be rewritten non-dimensionally as

$$\phi^{ss} = \ln (V_o/V_*), \quad \psi^{ss} = -(\beta - 1)\phi^{ss}. \tag{2.10}$$

This describes a straight line with slope $-(\beta - 1)$ in the (ψ, ϕ) plane. In particular, if V_* is chosen to equal $V_o \neq 0$, the steady state of the system is at the origin of the phase plane.

The critical spring stiffness for linear stability given by (2.6) can be written using (2.7) as

$$k_{cr} = \beta - 1 \tag{2.11}$$

and the velocity-weakening condition $B > A$ translates to

$$\beta > 1. \tag{2.12}$$

To commence the phase plane analysis, T is first eliminated from the governing equations (2.8) and (2.9) to get an equation of the form

$$P(\psi, \phi)d\psi + Q(\psi, \phi)d\phi = 0 \tag{2.13}$$

where

$$P(\psi, \phi) = k(v_0 - e^\phi) - \beta \left[e^{(\phi-\psi)/\beta} - e^\phi \right] \tag{2.14}$$

$$Q(\psi, \phi) = -k(v_0 - e^\phi). \tag{2.15}$$

13

An integrating factor of the form $e^{q(\psi,\phi)}$ is now sought such that

$$dU = [P(\psi,\phi)d\psi + Q(\psi,\phi)d\phi]e^{q(\psi,\phi)} \tag{2.16}$$

is a perfect differential. This requires that

$$\frac{\partial[e^{q(\psi,\phi)}P(\psi,\phi)]}{\partial\phi} = \frac{\partial[e^{q(\psi,\phi)}Q(\psi,\phi)]}{\partial\psi} \tag{2.17}$$

Substituting for P and Q and simplifying, we get

$$kv_o\left[\frac{\partial q}{\partial\psi} + \frac{\partial q}{\partial\phi}\right] - ke^\phi\left[\frac{\partial q}{\partial\psi} + \frac{\partial q}{\partial\phi} + 1\right] - \beta e^{(\phi-\psi)/\beta}\left[\frac{\partial q}{\partial\phi} + \frac{1}{\beta}\right] + \beta e^\phi\left[\frac{\partial q}{\partial\phi} + 1\right] = 0 \tag{2.18}$$

The most general solution of (2.18) for arbitrary k and v_o could not be found. However, solutions in which q is linear in its variables could be found when $k = k_{cr}$ with arbitrary v_o and when $v_o = 0$ with arbitrary k. We consider the two cases separately below:

2.2.1 Case 1: $k = k_{cr}$, arbitrary $v_o \neq 0$

First, the case $k = k_{cr} \equiv \beta - 1$ (*i.e* $K = K_{cr}$) with a non-stationary load point is studied. The phase plane trajectories obtained here are used in Section 2.3 to construct a Liapunov function and hence establish results on the stability of steady-state sliding for arbitrary perturbations from steady state.

When $k = k_{cr}$ and $v_o \neq 0$, it can be shown that

$$q = (\psi - \phi)/\beta \tag{2.19}$$

is a solution to (2.18). On integrating (2.16), the trajectories in phase plane are found to be given by

$$U = k\beta\left[v_o + \frac{e^\phi}{\beta - 1}\right]e^{(\psi-\phi)/\beta} - \psi\beta = \text{constant} \qquad ; \; k = k_{cr} \equiv \beta - 1 \tag{2.20}$$

2.2.2 Case 2: $v_o = 0$, arbitrary k

Explicit phase plane trajectories could also be determined for the case of a stationary load point, with the spring stiffness being arbitrary. This situation may be taken to model, for instance, the stressing of a stationary fault segment by a large earthquake in its vicinity. It is of interest to know whether the stress change associated with the earthquake can trigger a delayed instability (aftershock) in the fault segment. Such an instability mechanism is referred to as inducing a state of accelerating self-driven creep (see Rice and Gu (1983), Dieterich (1994)). The results obtained below will be used in Section 2.4 to derive a simple condition for the existence of such an instability.

14

When $v_o = 0$, it can be shown that

$$q = \frac{(\beta - 1)(\beta - k)}{k\beta}\psi - \frac{\phi}{\beta} \quad (2.21)$$

is a solution to (2.18). The trajectories, obtained by integrating (2.16), are

$$U = \left[\beta e^{\phi} e^{(\psi - \phi)/\beta} - \frac{(\beta - 1)\beta}{\beta - 1 - k}\right]\frac{k}{\beta - 1}e^{(\beta - 1 - k)\psi/k} = \text{constant} \quad ; k \neq k_{cr} \quad (2.22)$$

$$U = \beta e^{\phi} e^{(\psi - \phi)/\beta} - \psi\beta = \text{constant} \quad ; k = k_{cr} \quad (2.23)$$

2.3 Stability results with non-stationary load point

In this section, the phase plane trajectories obtained in Section 2.2.1 for the case of critical spring stiffness are used to study the stability of steady state slip when the load point is non-stationary (moving at constant velocity $V_o > 0$).

2.3.1 Case 1: $k = k_{cr}$

We show that slip motion is always periodic when $k = k_{cr}$ and the velocity weakening condition, $\beta > 1$ is satisfied. In other words, the phase plane trajectories (2.20) always form closed contours when $\beta > 1$. We establish this result by showing that for given values of U and ψ, there exist either two or no values of ϕ that satisfy (2.20) and, similarly, for specified values of U and ϕ, there exist either two or no values of ψ satisfying (2.20). First, choose $V_* = V_o$ (*i.e* $v_o = 1$) without loss of generality. For given values of U and ψ, (2.20) may be written in the form

$$e^{-\phi/\beta} + \frac{e^{\phi(\beta - 1)/\beta}}{\beta - 1} = \text{constant} \quad (2.24)$$

Now, from a graphical construction of the left and right hand sides of the above equation, it is easily seen that there are either two or no values of ϕ that satisfy the above equation when $\beta > 1$. Similarly, for specified values of U and ϕ, (2.20) gives

$$(\text{constant})e^{\psi/\beta} = \psi\beta + U \quad (2.25)$$

As before, it may shown that the above equation is satisfied by either two or no values of ψ. Hence, the trajectories form closed contours when $\beta > 1$ and slip is always periodic. A typical plot of the phase plane trajectories for this case is shown in Fig. 2.1, with $\beta = 5/4$.

This result may be contrasted with the one obtained by Gu et al. (1984) for the Ruina-Dieterich slip law. They show that, when $k = k_{cr}$ and $\beta > 1$, with friction evolving according to the slip law, there exist initial conditions leading to both periodic as well as

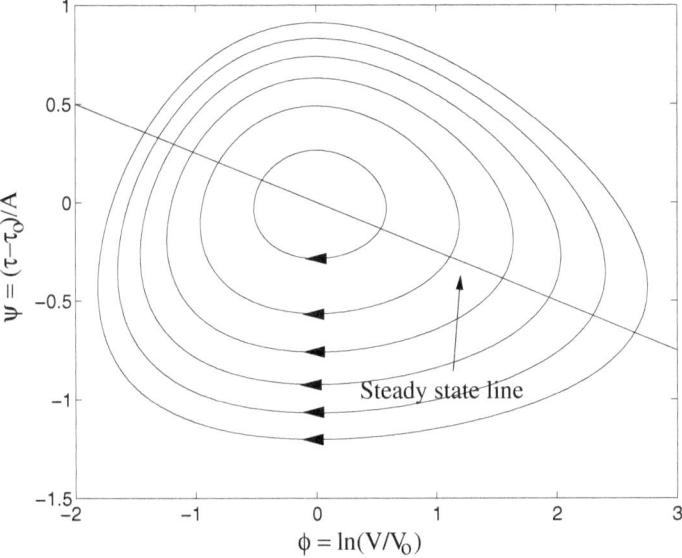

Figure 2.1: Trajectories in phase plane for slip motion under constant, non-zero imposed load-point velocity V_o, with $\beta = 5/4$, $k = k_{cr} = 1/4$. V_* has been chosen to equal V_o.

unstable slip motions. Perturbations that displace the system sufficiently away from the steady state solution cause the instability.

2.3.2 Case 2: $k \neq k_{cr}$

The stability of steady-state slip for generic $k \neq k_{cr}$ is determined by finding a Liapunov function for the problem. It is shown that with a velocity weakening Dieterich-Ruina ageing law, slip is always stable when $k > k_{cr}$ and always unstable when $k < k_{cr}$.

Consider the function generated by adding a certain function of ψ to U of (2.20):

$$U_1 = k\beta \left[v_o + \frac{e^\phi}{\beta - 1} \right] e^{(\psi - \phi)/\beta} - \psi\beta - v_o^{(\beta-1)/\beta}\beta^2 \left[\frac{k}{\beta - 1} - 1 \right] e^{\psi/\beta} \qquad (2.26)$$

It is easily established that U_1 is a Liapunov function when $\beta > 1$ since:

1. Trajectories of constant U_1, when they exist, are closed contours around the steady state solution (2.10) when $\beta > 1$. This may be shown by graphical constructions similar to the ones presented earlier. The global minimum of U_1 occurs at the steady state solution (2.10).

2. The derivative of U_1 along a solution trajectory is given by

$$\frac{dU_1}{dT} = \frac{\partial U_1}{\partial \psi}\frac{d\psi}{dT} + \frac{\partial U_1}{\partial \phi}\frac{d\phi}{dT}. \qquad (2.27)$$

Using (2.26) and the governing equations (2.8) and (2.9), this may be evaluated as

$$\frac{dU_1}{dT} = -k \left[\frac{k}{\beta - 1} - 1 \right] \beta e^{\psi/\beta}(e^\phi - v_o)(e^{(\beta-1)\phi/\beta} - v_o^{(\beta-1)/\beta}) \qquad (2.28)$$

Clearly, the factor $(e^\phi - v_o)(e^{(\beta-1)\phi/\beta} - v_o^{(\beta-1)/\beta})$ in (2.28) is always of positive sign when $\beta > 1$. Hence, when $k > \beta - 1(= k_{cr})$ and $\beta > 1$, $dU_1/dT < 0$ and every solution trajectory evolves with monotonically decreasing values of U_1. Since the global minimum of U_1 occurs at the steady state solution when $\beta > 1$, it follows that every initial condition evolves towards the steady-state solution. Therefore, all slip motions are stable when $k > k_{cr}$ and $\beta > 1$. Similarly, when $k < \beta - 1(= k_{cr})$ and $\beta > 1$, $dU_1/dT > 0$. Every initial condition evolves towards ever increasing values of U_1. Hence, all slip motions are unstable with slip velocity becoming unbounded.

These results may be compared with the analogous results obtained by Gu et al. (1984) for the Ruina-Dieterich slip law. They show that when $k > k_{cr}$, initial conditions corresponding to a sufficiently small perturbation from the steady state solution give rise to stable slip, while others cause unstable slip. When $k < k_{cr}$ slip is always unstable, as for the Dieterich-Ruina ageing law.

17

2.3.3 Discussion

The stability results obtained in this section have an important implication for nucleation of slip instabilities in continuum systems. Consider a patch of linear dimension D slipping frictionally in elastic surroundings with shear modulus μ. An effective stiffness can be identified in this case as

$$K \sim \mu/D \tag{2.29}$$

For the Dieterich-Ruina ageing law, we know from the present analysis that unstable slip occurs only when

$$K < K_{cr} \Longleftrightarrow D > D_{cr} \tag{2.30}$$

Hence, a critical nucleation size, D_{cr}, has to be attained before unstable slip can occur. On the other hand, for the Ruina-Dieterich slip law, such a nucleation size can be defined for linearized stability analysis but is not strictly defined in general since unstable slip can occur at any value of K if the initial condition is sufficiently perturbed from the steady state.

The results of this section may conveniently be visualized in terms of a stability diagram in the $K - \Delta$ plane, where Δ is a perturbation in stress or sliding velocity from steady state. We have seen that in this plane, the straight line $K = K_{cr}$ forms the stability boundary dividing regions of stable and unstable slip. The effect of inertia of the block on the stability boundary has been studied numerically in work the details of which are not reported here. An additional parameter is introduced into the problem by inclusion of inertia. This may be taken to be the ratio of an inertial to a frictional time scale:

$$r = \left(\frac{T/2\pi}{L/V_*}\right), \tag{2.31}$$

where T is the period of free vibrations of the spring-mass system. An interesting feature of the results is that for small values of r, the stability boundary initialy bends towards higher values of K from K_{cr} and then bends back towards values of $K < K_{cr}$. This implies that in response to finite perturbation, periodic stick-slip motion can occur even for values of $K > K_{cr}$, albeit in a very narrow range. For larger values of r, the stability boundary lies throughout in the region $K \leq K_{cr}$.

2.4   Stability results for stationary load point

In this section, the phase plane trajectories obtained in Section 2.2.2 are used to establish conditions for the existence of self-driven creep modes of instability. We shall show that the spring stiffness $K = K_{cr}$ plays a critical role in dividing types of response. This is remarkable because K_{cr} arose in the context of a linearized stability analysis of steady sliding, which is not a mode of response when the load point is stationary.

We observe that when $v_o = 0$, the only long time behaviors of the system, consistent with the governing equations (2.8) and (2.9) are

$$\phi \to \infty \quad \text{and} \quad \psi \to -\infty, \quad \text{and} \tag{2.32}$$

$$\phi \to -\infty \quad \text{and} \quad \psi \to -\infty.$$

We consider the two cases $k < k_{cr}$ and $k \geq k_{cr}$ separately below:

2.4.1   Case 1: $k < k_{cr}$

First, consider the case when $k < k_{cr}$. The trajectories corresponding to $U = 0$ are straight lines given by

$$\psi = -(\beta - 1)\phi + \beta \ln[(\beta - 1)/(\beta - 1 - k)]. \tag{2.33}$$

These are parallel to the steady state line in the (ϕ, ψ) plane, but located above it (at higher ψ, for given ϕ).

When $U > 0$, it follows from (2.22) that

$$e^{\phi}e^{(\psi-\phi)/\beta} > (\beta - 1)/(\beta - 1 - k). \tag{2.34}$$

Using this condition in (2.8), it can be shown that

$$d\phi/dT > ke^{\phi}/(\beta - 1). \tag{2.35}$$

Now, since $\int_{\phi}^{\infty} e^{-\phi'} d\phi'$ is bounded, $\phi \to \infty$ in finite time. Therefore, the slip velocity becomes unbounded in finite time when $U > 0$. The unboundedness of slip velocity in this case is due to the neglect of inertial effects.

From (2.22), we can write

$$e^{(\beta-1)\phi/\beta} = \left[\frac{U}{\beta k} e^{-(\beta-1-k)\psi/k} + \frac{1}{\beta - 1 - k} \right] (\beta - 1)e^{-\psi/\beta}. \tag{2.36}$$

When $U < 0$, with $k < \beta - 1 (= k_{cr})$, $\beta > 1$ and $\psi \to -\infty$ according to (2.32), it is clear that ϕ decreases at long times along every trajectory. Hence slip is stable when $U < 0$.

A typical plot of the phase plane trajectories when $k < k_{cr}$ is shown in Fig. 2.2. As has been shown in the analysis above, the straight line trajectory corresponding to $U = 0$ divides the phase plane into stable and unstable regions. Slip is stable when initial conditions cause $U < 0$ and unstable when $U > 0$. Gu et al. (1984) found a similar division of the phase plane in the stationary load-point case, but for their analysis, with the slip law (2.4), such division exists for all $k > 0$, and here it exists only for $k < k_{cr}$.

19

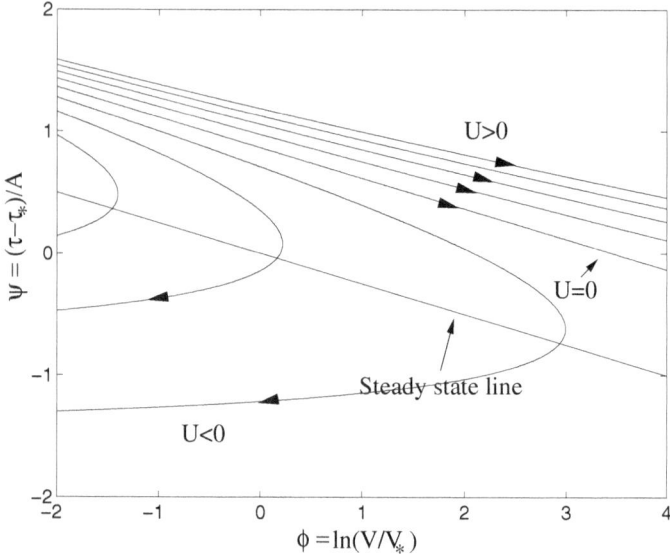

Figure 2.2: Trajectories in phase plane for slip motion with stationary load point and $\beta = 5/4$, $k = 1/8 < k_{cr}$. Self-driven creep occurs when $U > 0$.

2.4.2 Case 2: $k \geq k_{cr}$

Next, we show that when $k \geq k_{cr}$, slip is always stable. It is easily seen that, when $k \geq k_{cr}$ and $\beta > 1$, $\phi \to \infty$ is inconsistent with U being a constant along a trajectory. Hence, according to (2.32), the only long time behavior permissible is that $\phi \to -\infty$ $(i.e$ $V \to 0)$. In other words slip is always stable when $k \geq k_{cr}$. Typical plots of trajectories for $k = k_{cr}$ and $k > k_{cr}$ are shown in Figs. 2.3 and 2.4 respectively.

2.4.3 Discussion

It is clear from the above analysis that self-driven creep modes of instability can exist, when the ageing law (2.3) applies, only when $k < k_{cr}$. The amount of stress perturbation, $\Delta\psi$ required to drive a system from steady state to instability is the distance of the stability boundary (2.33) from the steady state line (2.10):

$$\Delta\psi = \beta \ln[(\beta - 1)/(\beta - 1 - k)]. \tag{2.37}$$

In contrast, for the Ruina-Dieterich slip law, Gu et al. (1984) show that stable and unstable regions exist for all $k_{cr} > 0$. Hence, self-driven creep modes exist for all values of k. In the unstable region, slip velocity becomes unbounded in finite time, as in the present study.

2.5 Conclusions

The stability of quasi-static frictional slip of a rigid block loaded by a linear spring has been studied. A rate- and state-dependent frictional constitutive law (2.2) with state evolution described by the Dieterich-Ruina ageing law (2.3) has been adopted in this study. A non-linear stability analysis of the steady state solution (2.5) using analytically determined phase-plane trajectories and Liapunov function techniques has been carried out. The stability results are shown to have an extremely simple form:

- With non-zero load-point velocities $(V_o \neq 0)$, slip motion is always periodic when

$$K = K_{cr} = (B - A)/L > 0.$$

When $K > K_{cr} > 0$, sliding is always stable. In other words, the block always approaches the velocity of the imposed motion. When $K < K_{cr}$, slip is always unstable with the sliding velocity becoming unbounded.

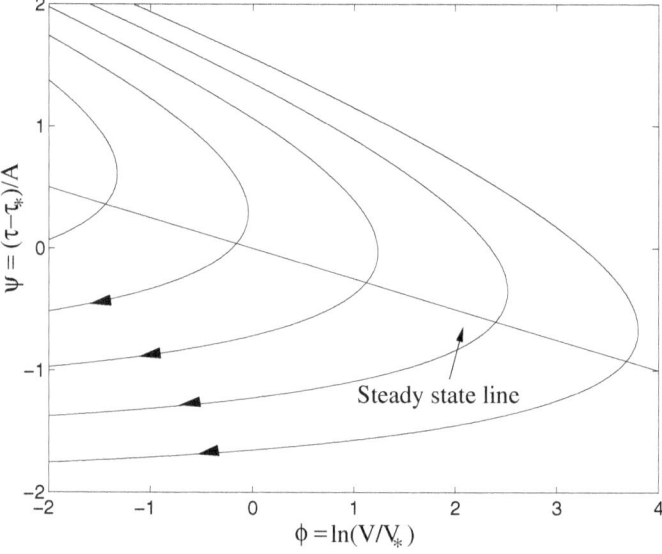

Figure 2.3: Trajectories in phase plane for slip motion with stationary load point and $\beta = 5/4$, $k = k_{cr} = 1/4$

22

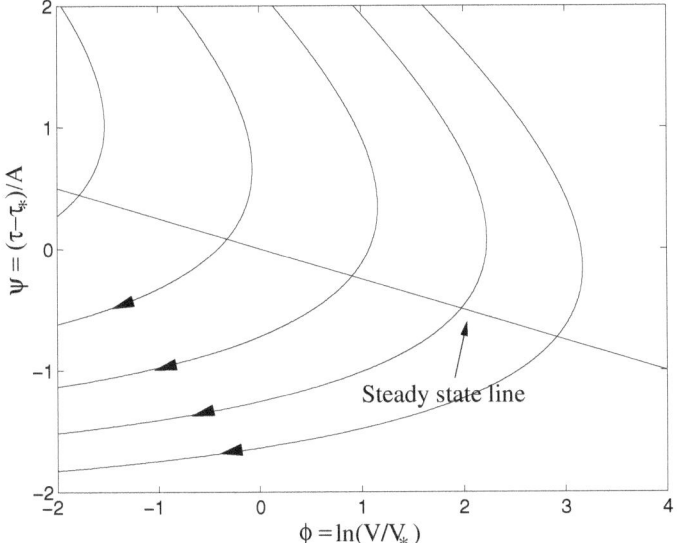

Figure 2.4: Trajectories in phase plane for slip motion with stationary load point and $\beta = 5/4$, $k = 1/2 > k_{cr}$

- When the load point is held stationary ($V_o = 0$), the system stably evolves towards ever-slower slip rates whenever $K \geq K_{cr} > 0$. However, when $K < K_{cr}$, initial conditions leading to stable and unstable slip motion exist. This shows that self-driven creep modes can exist only in the latter case. The unstable motions are shown to be such that the slip velocity becomes unbounded in finite time, corresponding to a delayed instability, or an aftershock.

The preclusion of instabilities when $K > K_{cr}$ has an important implication for slip instabilities in continuum systems. In a continuum system, a critical nucleation size of coherent slip has to be attained before unstable slip can ensue and the nucleation size does not depend on the strength of the perturbation.

Acknowledgements

This study was supported by ONR under grant N00014-96-10777 and USGS under grant 1434-HQ-97-GR-03094.

Chapter 3

Ill-posedness in steady sliding at a bimaterial interface

It has been shown recently that steady frictional sliding along an interface between dissimilar elastic solids with Coulomb friction acting at the interface is ill-posed for a wide range of material parameters and friction coefficients. The ill-posedness is manifest in the unstable growth of interfacial disturbances of all wavelengths, with growth rate inversely proportional to the wavelength. We establish the connection between the ill-posedness and the existence of a certain interfacial wave in frictionless contact, called the generalized Rayleigh wave. Precisely, it is shown that for material combinations where the generalized Rayleigh wave exists, steady sliding with Coulomb friction is ill-posed for arbitrarily small values of friction. In addition, intersonic unstable modes and supersonic steady-state modes exist for sufficiently large values of the friction coefficient.

3.1 Introduction

Recent work [Renardy (1992), Adams (1995), Martins et al. (1995), Martins and Simões (1995), Simões and Martins (1998)] has shown the ill-posedness, in the sense described below, of steady sliding of an elastic half-space against a dissimilar elastic half-space when Coulomb friction acts at their interface. Let V_o denote the velocity of steady sliding, the same at every point along the interface, and τ the shear stress at the interface. When the shear stress is perturbed in a single spatial mode of wavenumber k,

$$\Delta \tau = Q(t)e^{ikx_1}, \tag{3.1}$$

where x_1 is the coordinate axis along the interface and $Q(t)$ is an arbitrary function of time, t, propagating slip-rate modes of form

$$\Delta V = A(k)e^{ik(x_1-ct)}e^{a|k|t}$$

are found, $A(k)$ is the amplitude of the mode, a and c are independent of the wavelength, and where $a > 0$ for a broad range of friction coefficients and material pairs. For such $a > 0$ cases, all wavelengths in the slip response are unstable and the growth rate of the instability is inversely proportional to the wavelength. An observer traveling with the phase velocity c of the instability sees a perturbation velocity field that is the sum of an infinite number of modes, namely,

$$\Delta V(x+ct,t) = \int_{-\infty}^{+\infty} A(k)e^{ikx}e^{a|k|t}dk$$

where $x = x_1 - ct$. Clearly, this integral fails to exist (diverges by oscillation for $x \neq 0$) in an arbitrarily small time after the perturbation is turned on, unless $A(k)$ decays exponentially or faster with $|k|$. Such a problem is said to be ill-posed.

Renardy (1992) studied the sliding of a neo-Hookean elastic solid against a rigid substrate. In the limit of linear elasticity, he showed that ill-posedness exists when the friction coefficient is sufficiently high, namely, greater than unity. This limiting case was studied independently by Martins et al. (1995). On the other hand, Adams (1995) showed that when the two solids on either side of the interface are linear elastic and not very dissimilar, the problem can be ill-posed for arbitrarily small values of friction.

Earlier, Weertman (1963), Gol'dshtein (1967) and Achenbach and Epstein (1967) had shown that in frictionless sliding of dissimilar elastic half-spaces, constrained against formation of opening gaps, an interfacial wave solution can exist when the material mismatch is not very high. It is called the generalized Rayleigh wave since its speed of propagation reduces to that of the Rayleigh surface wave when the two materials are identical. The numerical results of Adams (1995) suggested a connection between the existence of the generalized Rayleigh wave and the ill-posedness. This is fully explored in our present work. We show that for conditions under which the generalized Rayleigh wave exists in frictionless contact, the stability problem with Coulomb friction is ill-posed for arbitrarily small values of friction.

Weertman (1980) argued that when such a wave exists, a self-healing slip pulse can propagate along the frictional interface between dissimilar elastic solids, even when the remote shear stress is less than the frictional strength of the interface, and a family of such pulse solutions has been constructed by Adams (1998). The velocity of propagation of the slip pulse is precisely that of the generalized Rayleigh wave. Numerical studies of the nucleation and propagation of such slip pulses with a Coulomb friction law at the interface, by Andrews and Ben-Zion (1997), Ben-Zion and Andrews (1998) and Harris and Day (1997) have difficulties that seem to have their origin in the instability and ill-posedness results cited earlier. They observe that their simulations depend on mesh size and that a nucleated slip pulse splits into a number of pulses. Using the spectral numerical methodology for bimaterials [Breitenfeld and Geubelle (1998)], Cochard and Rice (2000) illustrate the ill-posedness by showing that the more terms in their spectral basis set, the more the pulse splitting for a case which is ill-posed in the sense discussed above. They show that the same method gives results which converge with enlargement of the basis set for parameter choices in the well-posed range.

This chapter is organized as follows: In Section 3.2, the governing elastodynamic relations between perturbations (from steady sliding) in slip and opening and those in shear and normal stress at the interface are derived. These relations are used in subsequent sections and in the next chapter to derive the governing equations for stability. First, the case of frictionless sliding is reviewed in Section 3.3, leading to a discussion of the generalized Rayleigh wave of Weertman (1963), Gol'dshtein (1967) and Achenbach and Epstein (1967). In Section 3.4, slip stability with Coulomb friction at the interface is studied. A simpler rederivation of the ill-posedness results of Adams (1995) is presented. The connection between the existence of the generalized Rayleigh wave and the ill-posedness is explicitly shown. Numerical stability diagrams are constructed for a range of bimaterial parameters and friction coefficients of practical interest.

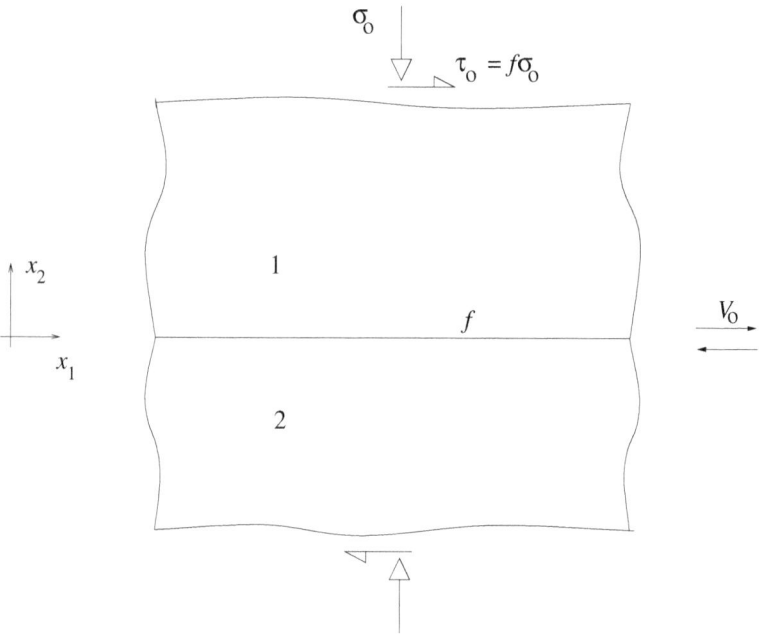

Figure 3.1: Frictional sliding along an interface between dissimilar materials

3.2 Elastodynamic relations

Let (x_1, x_2) be Cartesian coordinates defined such that the slip plane lies at $x_2 = 0$ (see Fig. 3.1) and the steady state velocities of the material points on either side of the interface, $x_2 = 0^+$ and $x_2 = 0^-$ are, respectively, $+V_o/2$ and $-V_o/2$. Far away from the interface, a shear stress τ_o equal to the frictional strength of the interface is applied, i.e. $\tau_o = f\sigma_o$, where f is the Coulomb friction coefficient and σ_o is the remote compressive normal stress. The shear and dilatational wave speeds of the material in the region $x_2 \geq 0^+$ are denoted by c_{s1} and c_{d1}, respectively. Corresponding wave speeds in the lower half space, $x_2 \leq 0^-$ are denoted by c_{s2} and c_{d2}, respectively. Similar notation is used for the densities, ρ_1 and ρ_2, shear moduli, μ_1 and μ_2 and Poisson's ratios, ν_1 and ν_2, of the two solids. Without loss of generality, we assume that $c_{s2} > c_{s1}$.

Let $\sigma_{\alpha\beta}(x_1, x_2, t)$ and $u_\alpha(x_1, x_2, t)$ denote the stress and displacement fields. In this section, we derive elastodynamic relations between perturbations in traction components of the stress on the interface,

$$\tau_\alpha(x_1, t) = \sigma_{2\alpha}(x_1, 0, t) \tag{3.2}$$

and those in displacement discontinuities,

$$\delta_\alpha(x_1, t) = u_\alpha(x_1, 0^+, t) - u_\alpha(x_1, 0^-, t) \tag{3.3}$$

at the interface. We restrict our study to the case where the slip velocity $V \ll c_{s1}$, which is generally the regime of interest.

Consider a perturbation field in a single spatial mode of the form

$$
\begin{aligned}
\tau_1(x_1, t) &= \tau_o + T_1(t)e^{ikx_1}, \\
\tau_2(x_1, t) &= -\sigma_o + T_2(t)e^{ikx_1}, \\
u_1^\pm(x_1, t) &= \pm V_o t/2 + U_1^\pm(t)e^{ikx_1}, \\
u_2^\pm(x_1, t) &= U_2^\pm(t)e^{ikx_1}.
\end{aligned}
\tag{3.4}
$$

Denoting the Laplace transform with respect to t by

$$\hat{g}(p) = \int_0^\infty e^{-pt}g(t)dt, \tag{3.5}$$

following Geubelle and Rice (1995), it can be shown that perturbations in displacements at $x_2 = 0^+$ are related to those in shear and normal stress by

$$
\left\{ \begin{array}{c} \hat{T}_1 \\ \hat{T}_2 \end{array} \right\} = \left[\begin{array}{cc} \hat{G}_{11} & \hat{G}_{12} \\ \hat{G}_{21} & \hat{G}_{22} \end{array} \right] \left\{ \begin{array}{c} \hat{U}_1^+ \\ \hat{U}_2^+ \end{array} \right\}
\tag{3.6}
$$

where

$$
\begin{aligned}
\hat{G}_{11}(p,k) &= -\mu_1\,|k|\,\frac{\alpha_{d1}(1-\alpha_{s1}^2)}{1-\alpha_{s1}\alpha_{d1}}, \\
\hat{G}_{22}(p,k) &= -\mu_1\,|k|\,\frac{\alpha_{s1}(1-\alpha_{s1}^2)}{1-\alpha_{s1}\alpha_{d1}}, \\
\hat{G}_{12}(p,k) &= -i\mu_1 k\left(2-\frac{1-\alpha_{s1}^2}{1-\alpha_{s1}\alpha_{d1}}\right) = -\hat{G}_{21}(p,k),
\end{aligned}
\tag{3.7}
$$

with

$$
\alpha_{s1}=\sqrt{1+s^2/c_{s1}^2}, \qquad \alpha_{d1}=\sqrt{1+s^2/c_{d1}^2} \qquad \text{and} \qquad s=p/|k|. \tag{3.8}
$$

To ensure bounded displacements in $x_2 \geq 0$, (see Geubelle and Rice (1995), eq. 14), we require that the real parts of α_{s1} and α_{d1} be non-negative in the physical domain $\mathrm{Re}(s) \geq 0$. We make the real parts non-negative for all s by defining branch cuts in the complex s-plane along $(-i\infty, -ic_{s1}]$ and $[ic_{s1}, +i\infty)$ for α_{s1} and, similarly, along $(-i\infty, -ic_{d1}]$ and $[+ic_{d1}, +i\infty)$ for α_{d1}.

Inverting (3.6) we get

$$
\left\{\begin{array}{c} \hat{U}_1^+ \\ \hat{U}_2^+ \end{array}\right\} =
\left[\begin{array}{cc} \hat{C}_{11}^+ & \hat{C}_{12}^+ \\ \hat{C}_{21}^+ & \hat{C}_{22}^+ \end{array}\right]
\left\{\begin{array}{c} \hat{T}_1 \\ \hat{T}_2 \end{array}\right\}
\tag{3.9}
$$

where

$$
\begin{aligned}
\hat{C}_{11}^+(p,k) &= -\frac{1}{\mu_1\,|k|}\,\frac{\alpha_{s1}(1-\alpha_{s1}^2)}{R_1(s)}, \\
\hat{C}_{22}^+(p,k) &= -\frac{1}{\mu_1\,|k|}\,\frac{\alpha_{d1}(1-\alpha_{s1}^2)}{R_1(s)}, \\
\hat{C}_{12}^+(p,k) &= -\frac{1}{i\mu_1 k}\,\frac{2\alpha_{s1}\alpha_{d1}-(1+\alpha_{s1}^2)}{R_1(s)} = -\hat{C}_{21}^+(p,k),
\end{aligned}
\tag{3.10}
$$

with

$$
R_1(s) = 4\alpha_{s1}\alpha_{d1} - (1+\alpha_{s1}^2)^2. \tag{3.11}
$$

The properties of the Rayleigh function $R_1(s)$ are discussed, for instance, in Achenbach (1973). The function has a double zero at $s=0$ and simple zeros at $s=\pm ic_{R1}$, where c_{R1} is the speed of propagation of the Rayleigh wave at a free surface of the material. It is important to note that the combination

$$
\hat{C}_{11}^+\hat{C}_{22}^+ - \hat{C}_{12}^+\hat{C}_{21}^+ = (1 \quad \alpha_{s1}\alpha_{d1})/\mu_1^2 k^2 R_1(s), \tag{3.12}
$$

and thus has simple poles at $s=\pm ic_{R1}$, and not double poles as may be naively expected.

30

Analogous equations at $x_2 = 0^-$ are

$$\left\{ \begin{array}{c} \hat{U}_1^- \\ \hat{U}_2^- \end{array} \right\} = \left[\begin{array}{cc} -\hat{C}_{11}^- & \hat{C}_{12}^- \\ \hat{C}_{21}^- & -\hat{C}_{22}^- \end{array} \right] \left\{ \begin{array}{c} \hat{T}_1 \\ \hat{T}_2 \end{array} \right\} \tag{3.13}$$

where

$$\begin{aligned}
\hat{C}_{11}^-(p,k) &= -\frac{1}{\mu_2 |k|} \frac{\alpha_{s2}(1 - \alpha_{s2}^2)}{R_2(s)}, \\
\hat{C}_{22}^-(p,k) &= -\frac{1}{\mu_2 |k|} \frac{\alpha_{d2}(1 - \alpha_{s2}^2)}{R_2(s)}, \\
\hat{C}_{12}^-(p,k) &= -\frac{1}{i\mu_2 k} \frac{2\alpha_{s2}\alpha_{d2} - (1 + \alpha_{s2}^2)}{R_2(s)} = -\hat{C}_{21}^-(p,k),
\end{aligned} \tag{3.14}$$

with

$$\alpha_{s2} = \sqrt{1 + s^2/c_{s2}^2}, \quad \alpha_{d2} = \sqrt{1 + s^2/c_{d2}^2} \quad \text{and} \quad R_2(s) = 4\alpha_{s2}\alpha_{d2} - (1 + \alpha_{s2}^2)^2. \tag{3.15}$$

Branch cuts for α_{s2} and α_{d2} are defined analogous to those for α_{s1} and α_{d1} to ensure bounded displacements in the half space $x_2 \leq 0$. The non-trivial roots of the Rayleigh function $R_2(s)$ are denoted by $s = \pm ic_{R2}$. Note that the combination $(\hat{C}_{11}^- \hat{C}_{22}^- - \hat{C}_{12}^- \hat{C}_{21}^-)$ only has simple poles at $s = \pm ic_{R2}$.

Writing the perturbed slip and opening as

$$\begin{aligned}
\delta_1(x_1, t) &= V_o t + D_1(t) e^{ikx_1}, \\
\delta_2(x_1, t) &= D_2(t) e^{ikx_1},
\end{aligned} \tag{3.16}$$

we can subtract (3.13) from (3.9) at low slip rates, $V \ll c_{s1}$, to get

$$\left\{ \begin{array}{c} \hat{D}_1 \\ \hat{D}_2 \end{array} \right\} = \left[\begin{array}{cc} \hat{K}_{11} & \hat{K}_{12} \\ \hat{K}_{21} & \hat{K}_{22} \end{array} \right] \left\{ \begin{array}{c} \hat{T}_1 \\ \hat{T}_2 \end{array} \right\} \tag{3.17}$$

where

$$\hat{K}_{11} = \hat{C}_{11}^+ + \hat{C}_{11}^-, \qquad \hat{K}_{22} = \hat{C}_{22}^+ + \hat{C}_{22}^-, \qquad \hat{K}_{12} = \hat{C}_{12}^+ - \hat{C}_{12}^- = -\hat{K}_{21}. \tag{3.18}$$

The functions $\hat{K}_{ij}$ have simple poles at $s = \pm ic_{R1}$ and at $s = \pm ic_{R2}$.

We will find it convenient to use the inverse of eq. (3.17), namely

$$\left\{ \begin{array}{c} \hat{T}_1 \\ \hat{T}_2 \end{array} \right\} = \left[\begin{array}{cc} \hat{M}_{11} & \hat{M}_{12} \\ \hat{M}_{21} & \hat{M}_{22} \end{array} \right] \left\{ \begin{array}{c} \hat{D}_1 \\ \hat{D}_2 \end{array} \right\} \tag{3.19}$$

where

$$\hat{M}_{11} = \hat{K}_{22}/D, \qquad \hat{M}_{22} = \hat{K}_{11}/D, \qquad \hat{M}_{12} = -\hat{K}_{12}/D = -\hat{M}_{21}, \tag{3.20}$$

31

with

$$D = \hat{K}_{11}\hat{K}_{22} - \hat{K}_{12}\hat{K}_{21}. \tag{3.21}$$

The combination $(\hat{K}_{11}\hat{K}_{22} - \hat{K}_{12}\hat{K}_{21})$ only has simple poles at $s = \pm ic_{R1}$ and $s = \pm ic_{R2}$, following our previous discussion of the poles of a similar combination of $\hat{C}_{ij}^{\pm}$. Thus, the Rayleigh poles in the numerators of the expressions for $\hat{M}_{ij}$ cancel those in the denominator.

Suppose that a perturbation in shear stress is applied at the sliding interface (e.g. due to a wave incident on the interface) of the form

$$\Delta\tau_1(t,k) = Q(t)e^{ikx_1}$$

such that $Q(t) = 0$ for $t < 0$ and $Q(t)$ is arbitrary for $t > 0$. We interpret this as an additional stress (to τ_∞) that would have been supported if the interface was constrained against further slip (than $V_o t$) or opening. The total change in shear and normal stresses at the interface is the sum of externally applied perturbation and those due to heterogeneous slip and opening, given by (3.19):

$$\left\{ \begin{array}{c} \hat{T}_1 \\ \hat{T}_2 \end{array} \right\} = \left[\begin{array}{cc} \hat{M}_{11} & \hat{M}_{12} \\ \hat{M}_{21} & \hat{M}_{22} \end{array} \right] \left\{ \begin{array}{c} \hat{D}_1 \\ \hat{D}_2 \end{array} \right\} + \left\{ \begin{array}{c} \hat{Q} \\ 0 \end{array} \right\} \tag{3.22}$$

where $\hat{Q}(p)$ is the Laplace transform of $Q(t)$. In the subsequent sections and in the next chapter, we determine the sliding response $\hat{D}_1$ due to the applied $\hat{Q}$, when various friction laws are operative at the interface.

3.3   Frictionless sliding: The generalized Rayleigh wave

First, consider the case when there is no friction at the interface. Setting $\hat{T}_1 = 0$ and $\hat{D}_2 = 0$ in (3.22), we get

$$\hat{D}_1 = -\frac{\hat{Q}}{\hat{M}_{11}}.$$

Using (3.20), we can rewrite this expression for the slip perturbation in terms of the $\hat{K}_{ij}$ as

$$\hat{D}_1 = -\frac{\hat{K}_{11}\hat{K}_{22} - \hat{K}_{12}\hat{K}_{21}}{\hat{K}_{22}}\hat{Q}. \tag{3.23}$$

As noted earlier, the Rayleigh poles in the denominator and numerator of the transfer function above cancel each other. The only poles of the transfer function are the roots of the equation

$$\hat{K}_{22}(s) = 0, \tag{3.24}$$

which is precisely the equation for the generalized Rayleigh wave of Weertman (1963), Gol'dshtein (1967) and Achenbach and Epstein (1967). Cancelling the Rayleigh poles in the numerator and denominator of the transfer function, the above equation can be written in the form

$$-\hat{K}_{22}(s)R_1(s)R_2(s)\mu_1 c_{s2}^2 |k|/s^2 = (\mu_1/\mu_2)\alpha_{d2}(s)R_1(s) + (c_{s2}/c_{s1})^2\alpha_{d1}(s)R_2(s) = 0. \quad (3.25)$$

Weertman (1963), Gol'dshtein (1967) and Achenbach and Epstein (1967) showed that when the two materials are not very dissimilar, the above equation has pure imaginary roots $s_o = \pm i c_{GR}$ corresponding to steady interfacial wave propagation. When $c_{R2} < c_{s1}$, such roots always exist and the speed of the wave c_{GR} is such that

$$\min(c_{R1}, c_{R2}) < c_{GR} < \max(c_{R1}, c_{R2}).$$

On the other hand, when $c_{R2} > c_{s1}$, roots exist when

$$\rho_2/\rho_1 < (c_{s1}/c_{s2})^4 \alpha_{d2}(ic_{s1})/\alpha_{d1}(ic_{s1})R_2(ic_{s1})$$

and the wave speed $c_{GR} < c_{s1}$. For modestly different densities and Poisson's ratios, the generalized Rayleigh wave exists for mismatches in shear wave speeds up to about 40 %.

3.4   Ill-posedness with Coulomb friction

In this section, a simpler rederivation of the ill-posedness results of Adams (1995) and Martins et al. (1995) is first presented. This is then used to establish the relationship between the ill-posedness and the existence of the generalized Rayleigh wave.

Taking $V > 0$, the Coulomb friction law is

$$\tau_1 = -f\tau_2.$$

As before, we study the stability of sliding response to perturbations from the steady state at velocity V. Solving for $\hat{D}_1$ in (3.22) with the constraints

$$\hat{T}_1 = -f\hat{T}_2, \qquad \hat{D}_2 = 0, \quad (3.26)$$

we get,

$$\hat{D}_1 = -\frac{\hat{Q}}{\hat{M}_{11} + f\hat{M}_{21}}.$$

Using (3.20), this becomes,

$$\hat{D}_1 = -\frac{\hat{K}_{11}\hat{K}_{22} - \hat{K}_{12}\hat{K}_{21}}{\hat{K}_{22} - f\hat{K}_{21}}\hat{Q}. \quad (3.27)$$

The poles in the numerator of the above expression corresponding to the Rayleigh waves of the two materials are again cancelled by identical poles in the denominator. For instability, we thus need that a root of the equation

$$\hat{K}_{22}(s) - f\hat{K}_{21}(s) = 0 \qquad (3.28)$$

have a positive real part. Recalling from (3.18), (3.10) and (3.14) that $\hat{K}_{22}$ has $1/|k|$ dependence on k and that $\hat{K}_{21}$ is proportional to $1/k$, the roots of the above equation depend only on the sign of k and not its magnitude.

The kernels $\hat{K}_{22}$ and $\hat{K}_{21}$ have the following properties:

$$\hat{K}_{22}(s,k) = \overline{\hat{K}_{22}(\bar{s},k)} = \hat{K}_{22}(-s,k) = \hat{K}_{22}(s,-k), \qquad (3.29)$$

$$\hat{K}_{21}(s,k) = -\overline{\hat{K}_{21}(\bar{s},k)} = \hat{K}_{21}(-s,k) = -\hat{K}_{21}(s,-k), \qquad (3.30)$$

where the overbar denotes complex conjugation. We see immediately that if s^* is a root of (3.28), so is $-s^*$. Also, if s^* and $-s^*$ are the roots for wavenumber k, $\overline{s^*}$ and $-\overline{s^*}$ are the roots for wavenumber $-k$. Writing

$$\mathrm{Re}(s^*) = a \qquad \text{and} \qquad -\mathrm{sign}(k)\mathrm{Im}(s^*) = c, \qquad (3.31)$$

the slip response to perturbations with wavenumber k or $-k$ has propagating modes of the form

$$\delta_1(x_1,t) \sim (e^{ik(x_1-ct)}e^{+a|k|t}, e^{ik(x_1+ct)}e^{-a|k|t}, e^{-ik(x_1-ct)}e^{+a|k|t}, e^{-ik(x_1+ct)}e^{-a|k|t}). \qquad (3.32)$$

The modes propagate in opposite directions with phase velocity c. When a root of (3.28) has a positive real part a, one mode grows with an exponent of $a|k|$, while the other decays at the same rate. Thus there is unstable growth in the modal response with the growth rate being faster for short wavelengths. Further, the growing modes associated with a given root propagate with a unique velocity c and in the same direction along the interface for k and $-k$. When $c > 0$, the unstable modes propagate in the positive x_1 direction and when $c < 0$, they travel in the negative x_1 direction. Note also from (3.29) and (3.30) that when a root exists with a real part a, then a root also exists with that same real part a when f is changed to $-f$.

3.4.1 Ill-posedness at arbitrarily small f when c_{GR} exists

Here, we prove that when the material properties on either side of the interface are such that a generalized Rayleigh wave exists in frictionless sliding, the stability problem with

Coulomb friction is ill-posed for arbitrarily small values of the friction coefficient. In Section 3.3, we saw that when the generalized Rayleigh wave exists, the equation $\hat{K}_{22}(s) = 0$ has two imaginary roots, $s_o = \pm i c_{GR}$. We now show using a perturbation analysis that when f is changed from zero, one root moves into the right half s-plane while the other moves into the left half plane. The numerical results presented in Section 3.4.2 are in agreement with this analysis.

For small values of f, we expect roots of $\hat{K}_{22} - f\hat{K}_{21}$ close to s_o. Using a perturbation expansion of the form

$$s = s_o + f s_1 + ... \tag{3.33}$$

for the roots in (3.28) and retaining $O(f)$ terms, we get

$$s_1 = \hat{K}_{21}(s_o)/\hat{K}'_{22}(s_o). \tag{3.34}$$

Since $\hat{K}_{21}(s_o)$ and $\hat{K}'_{22}(s_o)$ are purely imaginary, s_1 is a real number. Furthermore, it follows from (3.29) and (3.30) that

$$
\begin{aligned}
\hat{K}_{21}(\pm i c_{GR}) &= \hat{K}_{21}(+i c_{GR}), \\
\hat{K}'_{22}(\pm i c_{GR}) &= \pm \hat{K}'_{22}(+i c_{GR}).
\end{aligned}
$$

Therefore, the term $f s_1$ in the perturbation expansion is of opposite signs (and real) for the two values $s_o = \pm i c_{GR}$. The roots move from $s_o = \pm i c_{GR}$ parallel to the $\text{Re}(s)$ axis as f is changed from zero. One root moves into the right half s-plane, while the other moves to the left. This makes the problem ill-posed for arbitrarily small friction for cases where the generalized Rayleigh wave exists in frictionless contact. Note in particular that the argument would apply for positive or negative f, both giving ill-posedness.

Suppose now that the two solids are sufficiently different such that the generalized Rayleigh wave does not exist in frictionless contact. This means that the equation $\hat{K}_{22}(s) = 0$ has no roots, not only along the imaginary axis but in the whole complex plane. The latter is easily seen as follows. From the discussion of roots given in the previous subsection, if there exists any root with non-zero $\text{Re}(s)$, in the response to an e^{ikx_1} perturbation, then there must exist a root with $\text{Re}(s) > 0$, hence showing growth in time. But this violates energy conservation, which must apply in the frictionless case, as argued below. The two sliding bodies have a strain energy corresponding to uniform stressing in the unperturbed configuration, and have a kinetic energy corresponding to the rigid translation at rate V. An e^{ikx_1} perturbation field then provides a positive definite change in both strain energy and kinetic energy (the cross terms of the respective quadratic forms for strain and kinetic energy

35

density integrate to zero). But such energy per wavelength cannot be greater (or smaller) than the energy put into the initial e^{ikx_1} perturbation. Hence no root to $\hat{K}_{22}(s) = 0$ with $\text{Re}(s) \neq 0$ can exist. (The same energy conservation argument does not apply, of course, when $f \neq 0$.)

Thus, if no generalized Rayleigh wave exists, $\hat{K}_{22}(s) = 0$ has no roots. If we now consider solutions to the problem with friction, $\hat{K}_{22}(s) - f\hat{K}_{21}(s) = 0$, it is clear that for arbitrarily small but non-zero f, there will be no roots. (Indeed, there will be none for sufficiently small positive or negative f). The possible exception is that a root might emerge at ∞ as f is altered slightly from zero, but this possibility is precluded by the expressions in eqs. (3.10) and (3.14) which show that $\hat{C}_{21}^{\pm}(s)/\hat{C}_{22}^{\pm}(s)$, and hence $\hat{K}_{21}(s)/\hat{K}_{22}(s)$, approach zero as $s \rightarrow \infty$. Thus, when no generalized Rayleigh wave exists, there will be an interval around $f = 0$, say $-f_c < f < f_c$ (where $f_c > 0$) for which there is stable response to an e^{ikx_1} perturbation. Note in particular that the analysis for this case predicts stability even in a range of negative friction coefficients, $-f_c < f < 0$.

3.4.2 Discussion of the roots of $\hat{K}_{22} - f\hat{K}_{21} = 0$

First, we show that for sufficiently large values of f, a family of steady-state (i.e., non-growing) supersonic interfacial wave solutions exist. Along the imaginary axis $s = i\zeta$, for $|\zeta| \geq \max(c_{d1}, c_{d2})$, both $\hat{K}_{22}(s)$ and $\hat{K}_{21}(s)$ are pure imaginary. Hence, $f = \hat{K}_{22}/\hat{K}_{21}$ is a real number and $s = i\zeta$ is a root corresponding to that value of f. These roots correspond to supersonic waves at the interface. They generalize the ones found by Adams (2000) for a bimaterial system with a rigid substrate. Adams' results can be obtained by letting $c_{s2}/c_{s1} \rightarrow \infty$ and $\mu_1/\mu_2 \rightarrow 0$ in the expressions (3.18) for $\hat{K}_{22}(s)$ and $\hat{K}_{21}(s)$ and using them in (3.28). This gives

$$f = \frac{ik}{|k|} \frac{\alpha_{d1}(1 - \alpha_{s1}^2)}{2\alpha_{s1}\alpha_{d1} - (1 + \alpha_{s1}^2)}. \tag{3.35}$$

Clearly, the right hand side is a real number when $s = i\zeta$ is purely imaginary with $|\zeta| \geq c_{d1}$.

To numerically compute the roots of (3.28) in the complex s-plane, we use the property that since f is a real number, $\text{Im}(\hat{K}_{22}/\hat{K}_{21}) = 0$ at any root location. The following steps are performed:

1) Fix a value of $\text{Re}(s) = r$.

2) Determine all values of s with $\text{Re}(s) = r$ for which $\text{Im}(\hat{K}_{22}/\hat{K}_{21}) = 0$.

3) Compute $f = \hat{K}_{22}/\hat{K}_{21}$. Ignore the root if $f < 0$.

The case studied in Figs. 3.2 and 3.3 is one where there is a modest mismatch in material properties across the interface such that the generalized Rayleigh wave exists in

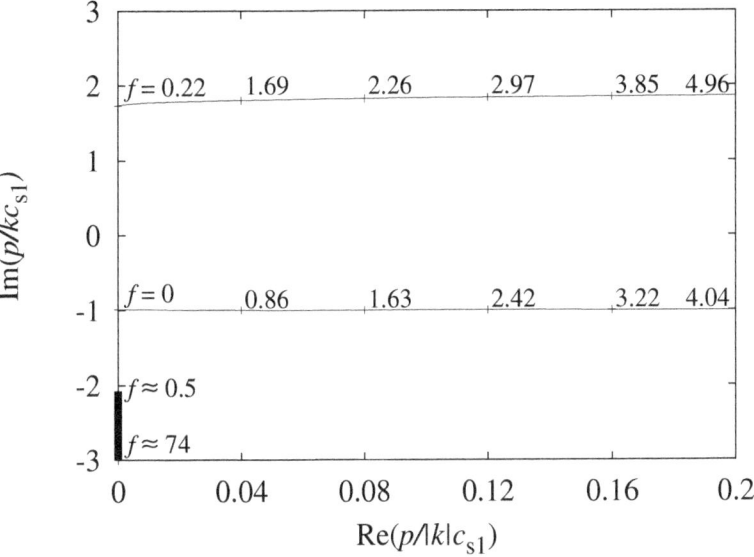

Figure 3.2: Root locations of $\hat{K}_{22}(s) - f\hat{K}_{21}(s) = 0$ in the complex right half s-plane as a function of friction coefficient f for a particular bimaterial pair: $c_{s2}/c_{s1} = 1.2$, $\rho_2/\rho_1 = 1.2$, and $\nu_1 = \nu_2 = 0.25$. The generalized Rayleigh wave exists in frictionless contact for this material combination and its speed is $c_{GR} = 0.9898c_{s1}$. Following the analysis in Section 3.4.1, roots exist in the right half s-plane for arbitrarily small f.

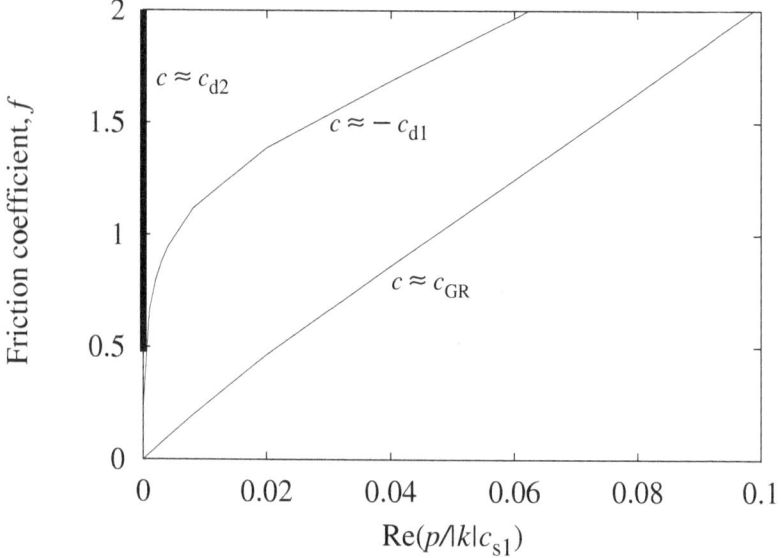

Figure 3.3: Normalized growth rate of instability as a function of friction coefficient for a particular bimaterial pair: $c_{s2}/c_{s1} = 1.2$, $\rho_2/\rho_1 = 1.2$, and $\nu_1 = \nu_2 = 0.25$. The generalized Rayleigh wave exists in frictionless contact for this material combination. Following the analysis of Section 3.4.1, instability occurs for arbitrarily small f. As defined in (3.31), $c = -\text{Im}(p/k)$ is the phase velocity of the instability. The speed of the supersonic modes is close to c_{d2} for the range of friction coefficients shown in the figure.

frictionless contact. The material properties are $c_{s2}/c_{s1} = 1.2, \rho_2/\rho_1 = 1.2, \nu_1 = \nu_2 = 0.25$. For this pair, the speed of the generalized Rayleigh wave is $c_{GR} = 0.9898c_{s1}$. We focus only on the roots in the right half s-plane since they determine stability. When $f = 0$, there are two roots at $s_o = \pm ic_{GR}$. As f is increased from zero, one of these roots moves into the right half s-plane, as seen in Figs. 3.2 and 3.3. Furthermore, Fig. 3.2 shows that the imaginary part of the root remains approximately constant as f is increased from zero. This implies that the velocity of propagation of the mode is approximately independent of the coefficient of friction in this regime (and thus $c \approx c_{GR}$, where c is defined as in (3.31)). These results are consistent with the perturbation analysis presented in the previous section.

For sufficiently high friction, other unstable modes are introduced. At $f = 0.22$, two roots appear on the branch cut of $\hat{K}_{22}(s)$ on the imaginary axis. One of these moves into the right half s-plane as f is further increased. Again, the propagation speed along the root path varies only modestly as f is increased, although for both roots, the rate of growth of the instability increases rapidly with f (Fig. 3.3). In addition to these unstable modes are the family of steady-state supersonic solutions mentioned earlier. They correspond to purely imaginary roots in Figs. 3.2 and 3.3 indicated with the thick line.

We next study in Figs. 3.4 and 3.5 a material pair with large contrast in properties across the interface so that the generalized Rayleigh wave does not exist in frictionless contact. The properties chosen are for a Steel/PMMA bimaterial system: $c_{s2}/c_{s1} = 2.40, \rho_2/\rho_1 = 6.58, \nu_1 = 0.35, \nu_2 = 0.3$. In this case, no roots exist in the right half s-plane when $f < 0.03$. Hence frictional sliding is stable in this range. At $f = f_c = 0.03$, a root appears on the right bank of the branch cut of $\hat{K}_{22}(s) - f\hat{K}_{21}(s)$, with propagation speed near c_{d1} and moves into the right half s-plane as f is further increased. Other unstable modes are also introduced at $f = 0.98$ and at the rather uninterestingly large $f = 6.31$. In addition, a family of supersonic steady-state solutions also exist at large friction.

Stability diagrams showing the range of values for which the stability problem is ill-posed can be constructed by calculating the value of the friction coefficient f_c at which an unstable right half pole first appears for a particular material combination. Two such diagrams are shown in Figs. 3.6 and 3.7, as a function of the shear wave speed mismatch. Fig. 3.6 is for the case when there is a modest 20% mismatch in densities between two Poisson materials ($\nu_1 = \nu_2 = 0.25$). The figure shows that when the shear wave speeds of the two materials are not too dissimilar so that the generalized Rayleigh wave exists in frictionless contact, $f_c = 0$ and the stability problem is ill-posed for arbitrary small values of friction. For slightly larger mismatch in shear wave speeds than in the previous case, an unstable mode propagates in the positive x_1 direction with a speed in the range (1 to

39

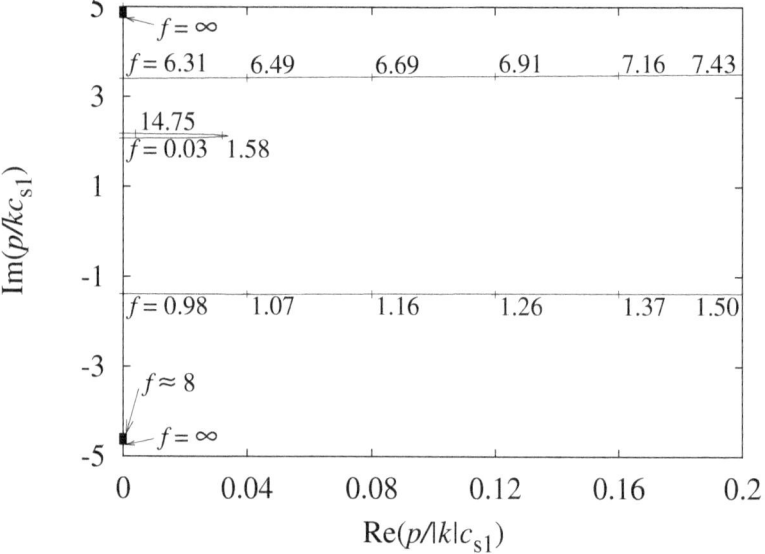

Figure 3.4: Root locations of $\hat{K}_{22}(s) - f\hat{K}_{21}(s) = 0$ in the complex right half s-plane as a function of friction coefficient f for Steel/PMMA material pair: $c_{s2}/c_{s1} = 2.40$, $\rho_2/\rho_1 = 6.58$, $\nu_1 = 0.35$ and $\nu_2 = 0.3$. The generalized Rayleigh wave does not exist in frictionless contact for this material combination. Following the analysis in Section 3.4.1, no roots exist in the complex s-plane for $|f| < f_c = 0.03$.

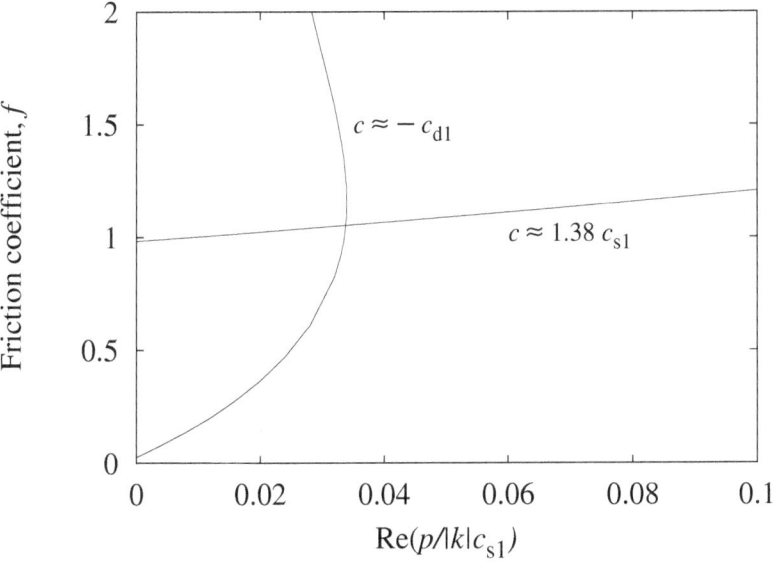

Figure 3.5: Normalized growth rate of instability as a function of friction coefficient for Steel/PMMA material pair: $c_{s2}/c_{s1} = 2.40$, $\rho_2/\rho_1 = 6.58$, $\nu_1 = 0.35$ and $\nu_2 = 0.3$. The generalized Rayleigh wave does not exist in frictionless contact for this material pair. Following the discussion in Section 3.4.1, instability occurs only when $|f| > f_c = 0.03$. As defined in (3.31), $c = -\text{Im}(p/k)$ is the phase velocity of the instability.

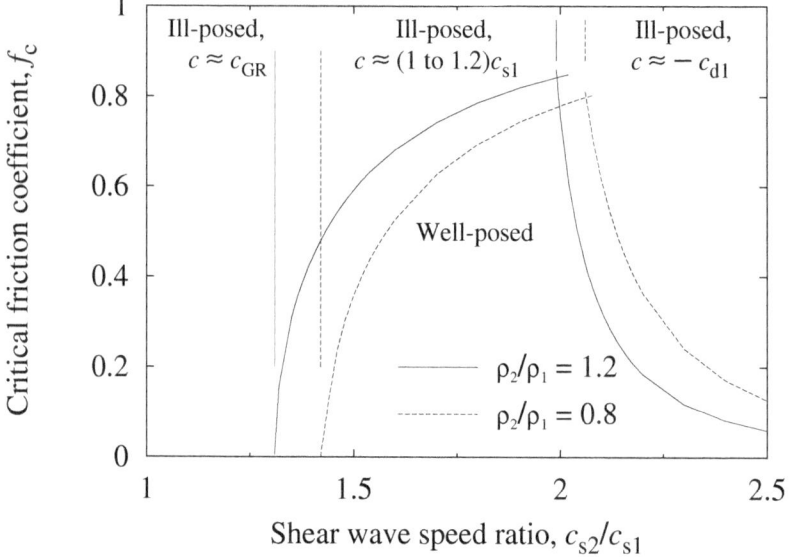

Figure 3.6: Stability diagram for two families of bimaterial pairs as a function of the shear wave speed ratio. For one pair, the density ratio is $\rho_2/\rho_1 = 1.2$ and for the other $\rho_2/\rho_1 = 0.8$. For both families of material pairs, $\nu_1 = \nu_2 = 0.25$. As defined in (3.31), $c = -\text{Im}(p/k)$ is the phase velocity of the instability.

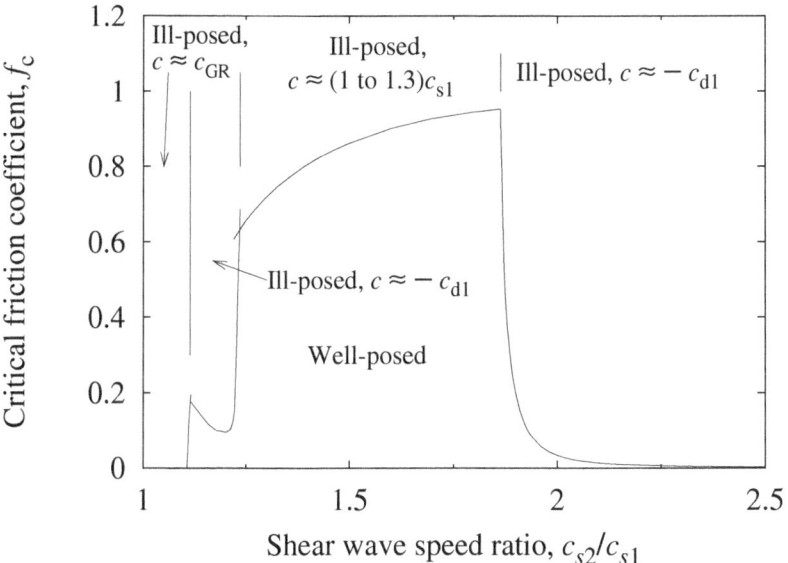

Figure 3.7: Stability diagram for a family of bimaterial pairs with $\rho_2/\rho_1 = 5.0$, $\nu_1 = 0.25$ and $\nu_2 = 0.35$ as a function of the shear wave speed ratio. $c = -\mathrm{Im}(p/k)$, as defined in (3.31), is the phase velocity of the instability.

1.2)c_{s1}. For still higher values of c_{s2}/c_{s1}, the destabilizing mode propagates at a speed very close to, but slightly lower than c_{d1} in the negative x_1 direction. Another stability diagram is shown in Fig. 3.7 for material pairs where $\rho_2/\rho_1 = 5.0, \nu_1 = 0.25$ and $\nu_2 = 0.35$, which are properties typical of a metal/carbon fiber composite bimaterial system (approximating the latter as isotropic). Here, as before, $f_c = 0$ when the shear wave speed mismatch is small enough to allow the existence of the generalized Rayleigh wave. When the generalized Rayleigh wave ceases to exist, an unstable mode propagates with a speed in the range (1 to 1.3)c_{s1}. For a narrow band of wave speed ratios the first destabilizing mode propagates at a speed close to and slightly higher than c_{d1}. This mode propagates in the negative x_1 direction. Again, at large values of c_{s2}/c_{s1}, an unstable mode is introduced at small values of friction that travels with a speed slightly lower than c_{d1} in the negative x_1 direction.

Acknowledgements

The study was supported by the Office of Naval Research through grant N00014-96-10777 to Harvard University, and through a Blaise Pascal Chair of the Foundation of Ecole Normale Superieure, Paris, to JRR. We are grateful to Drs. G. Adams and A. Cochard for discussions.

Chapter 4

Regularization of steady sliding at a bimaterial interface

Two experimentally motivated corrections to the Coulomb law are identified that regularize the problem of dynamic stability to perturbations from steady sliding at a dissimilar material interface. First, we show that a friction law incorporating a memory dependence, rather than instantaneous dependence, on normal stress gives rise to a well-posed stability problem. Such a dependence is suggested by high speed sliding experiments (at speeds of the order of 1 m/sec) on cutting tool materials. Secondly, we show that a friction law with a positive instantaneous logarithmic dependence on sliding velocity also regularizes the stability problem at sufficiently low sliding speeds. Experiments at speeds of up to a few mm/sec on a wide range of materials confirm such a dependence on sliding speed, as do theoretical models based on thermally activated creep at the asperity contacts.

4.1　Introduction

We study the dynamic stability to perturbations from the state of steady sliding depicted in Fig. 3.1 building on the framework developed in Chapter 3. At steady state, the sliding velocity is V_o and the remote shear and (compressive) normal stresses, τ_o and σ_o, respectively, are at the friction threshold, $\tau_o = f\sigma_o$, and equal, respectively, the shear and normal stresses at the interface. It was shown in Chapter 3 that a Coulomb friction law quite often leads to pathologies. Therefore, other features of friction, as suggested by experiments, are required to be incorporated into the friction law.

We discussed in Section 1.2 the corrections to the Coulomb law as motivated by experiments. It was argued that low speed sliding experiments suggest a linearized friction law for perturbations (from the steady state) in sliding velocity and normal stress can be written in the form

$$\tau^s - \tau_o = \frac{a\sigma_o}{V_o}(V - V_o) - \frac{b\sigma_o}{V_o} \int_0^t h_V(t')\dot{V}(t - t')dt' \tag{4.1}$$
$$+ (f - \alpha)(\sigma - \sigma_o) + \alpha \int_0^t h_\sigma(t')\dot{\sigma}(t - t')dt'$$

where $h_V(0) = h_\sigma(0) = 0$ and $h_V(t)$ and $h_\sigma(t) \to 1$ as $t \to \infty$. From the observed positivity of the direct effect of sliding velocity, it follows that $a > 0$.

High speed sliding experiments, on the other hand, were shown to have a different response to normal stress changes. As we discussed in Section 1.2, the experiments of Prakash and Clifton (1993) and Prakash (1998) suggest that there is no instantaneous change of shear strength, but rather a gradual change which occurs over a few microns of sliding, when a sudden change is made in the normal stress. This motivates a linear friction law of the form

$$\tau^s - \tau_o = f \int_0^t h_\sigma(t')\dot{\sigma}(t - t')dt' \tag{4.2}$$

where, as before, the function $h_\sigma(t)$ is chosen such that $h_\sigma(t) \to 1$ as $t \to \infty$.

In this chapter, we show that both constitutive forms (4.1) and (4.2) above give rise to mathematically well-posed stability problems, the former for sufficiently low V_o.

4.2　Regularization due to memory of normal stress

In this section, we study stability to perturbations with a simplified form of a friction law suggested by Prakash and Clifton (1993) and Prakash (1998). We also discuss a regularizing friction law proposed by Martins and Simões (1995) and Simões and Martins (1998) involving non-local spatial dependence of friction on normal stress.

4.2.1 Stability analysis with the simplified Prakash-Clifton law

As mentioned, the high speed sliding experiments of Prakash and Clifton (1993) and Prakash (1998) indicate that normal stress only has a memory effect on friction and has no instantaneous effect. They analyze their experiments in the framework of rate and state dependent friction, in which shear strength is regarded as a property of the current population of asperity contacts, and of their lifetimes, and it is only with ongoing slip or time that the population, and therefore the shear strength, can be altered. In a constitutive law proposed by them, the strength is assumed to be altered by the slip but not directly by the time since a normal stress change, although such remains to be verified experimentally. We use a simplification of their constitutive form which retains its essential feature for our purposes, namely, that there is a simple monotonic memory dependence but no instantaneous dependence of shear strength τ^s on compressive normal stress σ. The form used is:

$$\tau_1 = \text{sign}(V)\tau^s \text{ if } V \neq 0; \qquad \dot{\tau}^s = -(|V|/L)(\tau^s - f\sigma) \qquad (4.3)$$

where $\tau^s > 0$, $|\tau_1| \leq \tau^s$ if $V = 0$, and we are assuming $\sigma = -\tau_2 > 0$; V is the sliding velocity and $L > 0$ is a characteristic slip length over which the changes occur. No conclusion concerning regularization in the following development would change if we took L proportional to $|V|$, so that evolution of strength with time, rather than with slip, was described, or if we simply replaced $|V|/L$ with an expression of form $a + b|V| > 0$ where $a \geq 0$ and $b \geq 0$. We assume $V > 0$ below so that (4.3) requires

$$\dot{\tau}_1 = -(V/L)(\tau_1 + f\tau_2). \qquad (4.4)$$

Linearizing the above equation about the unperturbed slip rate V_o, taking Laplace transform, and considering a single Fourier mode as before, we get

$$p\hat{T}_1 = -(V_o/L)(\hat{T}_1 + f\hat{T}_2). \qquad (4.5)$$

Using this relation in (3.22), imposing the constraint $\hat{D}_2 = 0$ and solving for $\hat{D}_1$, we get

$$\hat{D}_1 = \frac{(sq+1)\hat{Q}}{(sq+1)\hat{M}_{11} + f\hat{M}_{21}}$$

where $q = L|k|/V_o$, or in terms of the $\hat{K}_{ij}$ as

$$\hat{D}_1 = \frac{(sq+1)(\hat{K}_{11}\hat{K}_{22} - \hat{K}_{12}\hat{K}_{21})}{(sq+1)\hat{K}_{22} - f\hat{K}_{21}}\hat{Q}. \qquad (4.6)$$

Thus, the equation governing stability is

$$(sq + 1)\hat{K}_{22}(s) - f\hat{K}_{21}(s) = 0. \tag{4.7}$$

We immediately see that no steady-state (i.e., s pure imaginary) supersonic solutions of (4.7) can now exist for $k \neq 0$ since both $\hat{K}_{22}(s)$ and $\hat{K}_{21}(s)$ are purely imaginary in that region. In the long wavelength limit, $|k| \to 0$ and hence $q \to 0$, the above equation reduces to (3.28), the governing equation for stability with constant Coulomb friction. Since ill-posedness relates to response as $|k| \to \infty$, it is of interest to know if there is stability at short wavelengths. This will require that the unstable roots of (3.28) move into left half plane or into a different Riemann sheet as $|k|$ is increased from 0 to k_{cr}.

In the limit $|k| \to \infty$, equivalent to $q \to \infty$, the equation determining stability becomes $\hat{K}_{22}(s) = 0$, which is precisely the condition for existence of the generalized Rayleigh wave in frictionless contact. Thus, if the material pair is sufficiently different that a generalized Rayleigh wave does not exist, which means that $\hat{K}_{22}(s) = 0$ has no solution, then we are assured of stability at sufficiently large $|k|$, $|k| > k_{cr}$. For cases where the generalized Rayleigh wave exists, a perturbation expansion in powers of $1/|k|$ for the roots gives the root location at large $|k|$ as

$$s = p/|k| = s_o + is_1/|k| + (s_2 + is_3)/k^2 + ...,$$

where $s_o = \pm ic_{GR}$ and s_1, s_2 and s_3 are real numbers with $s_2 > 0$ for one of the values of s_o. Therefore, a perturbation with large wavenumber k grows as $e^{s_2 t/|k|}$. This assures a finite integral over the amplitudes of all excited modes at all times and thus regularizes the problem. We thus see that for conditions under which the generalized Rayleigh wave exists in frictionless contact, all wavelengths are unstable with the friction law (4.3) as it was with the Coulomb friction law. However, the stability problem is now well-posed.

Generically, we find multiple bands of wavelengths that are unstable. We are concerned here only with determining the wavelength (or wavenumber k_{cr}) at which the first unstable mode that causes the pathological ill-posedness is regularized. This mode is stabilized at short wavelengths when the material parameters are such that generalized Rayleigh wave does not exist in frictionless sliding. Particular cases are illustrated in Figs. 4.1, 4.2 and 4.3. Fig. 4.1 is for a Steel/PMMA material pair. It was shown earlier in Fig. 3.5 that there are at most two unstable modes in the long wavelength limit when f is between 0 and 2. The critical wavenumber k_{cr} above which the first unstable mode is stabilized is shown in Fig. 4.1.

Fig. 4.2 shows a similar plot for the case of a material pair with $\rho_2/\rho_1 = 1.2$ and $\nu_1 = \nu_2 = 0.25$. We saw in Fig. 3.6 that the critical friction coefficient for this case is

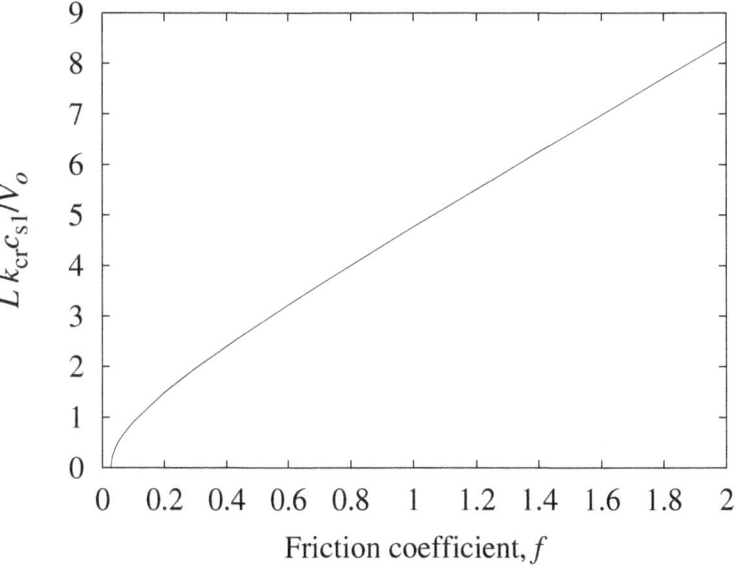

Figure 4.1: Regularization of short wavelength instability with simplified Prakash-Clifton friction law for Steel/PMMA material pair where $c_{s2}/c_{s1} = 2.40$, $\rho_2/\rho_1 = 6.58$, $\nu_1 = 0.35$ and $\nu_2 = 0.3$

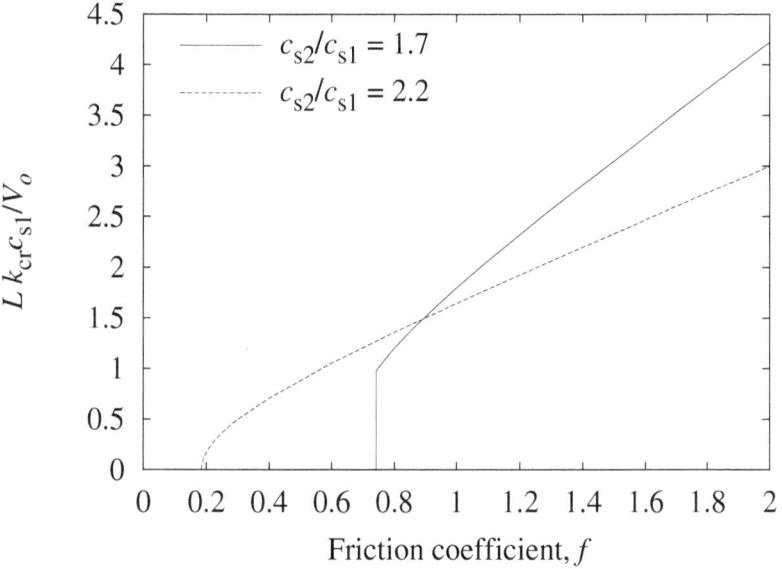

Figure 4.2: Regularization of short wavelength instability with simplified Prakash-Clifton friction law for bimaterial pairs with $\rho_2/\rho_1 = 1.2$, $\nu_1 = \nu_2 = 0.25$.

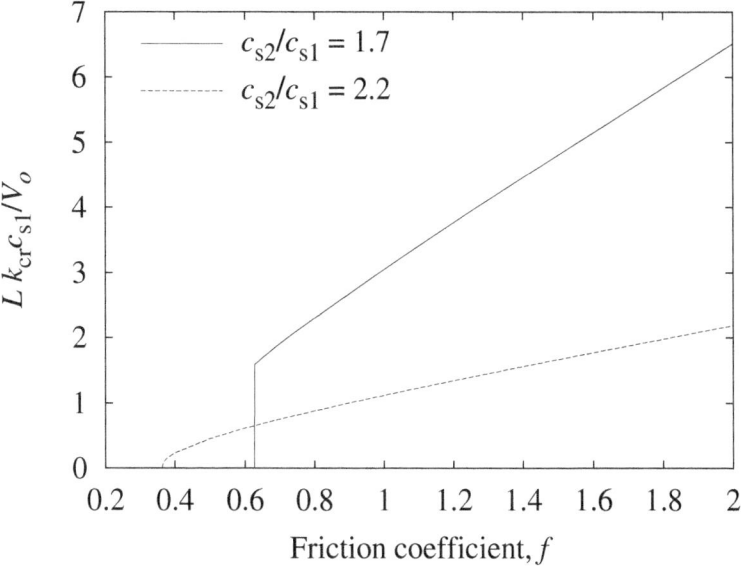

Figure 4.3: Regularization of short wavelength instability with simplified Prakash-Clifton friction law for bimaterial pairs with $\rho_2/\rho_1 = 0.8$, $\nu_1 = \nu_2 = 0.25$

$f_c = 0.742$ when $c_{s2}/c_{s1} = 1.7$ and $f_c = 0.184$ when $c_{s2}/c_{s1} = 2.2$. Now, Fig. 4.2 shows the wavelength at which the first unstable mode is stabilized for these two cases. Fig. 4.3 is a similar plot for an interface with $\rho_2/\rho_1 = 0.8$ and $\nu_1 = \nu_2 = 0.25$ at the same two values of c_{s2}/c_{s1} as in the previous case.

4.2.2 Further discussion of regularization

Previous studies by Martins and Simões (1995) and Simões and Martins (1998) showed that regularization can also be achieved by using a friction law in which the usual instantaneous dependence on normal stress is replaced by a dependence on normal stress averaged over some small finite area, although such a law seems not to be directly motivated by experiments We explain briefly how their regularization scheme can be analyzed in the present framework.

For simplicity, we consider the context of friction without any memory dependence so that, for slip situations which can be reduced to 2D perturbations like here, we write the shear strength as

$$\tau^s(x_1, t) = f \int_{x-D}^{x+D} w(|x_1' - x_1|)\sigma(x_1', t)dx_1' \qquad (4.8)$$

where the weight $w(|x_1|) \geq 0$ and $\int_{-D}^{+D} w(|x_1'|)dx_1' = 1$. As before, $\tau_1 = \tau^s$ when $V > 0$. Perturbations in modal shear and normal stresses of the form (3.4) at the interface are then related by

$$T_1(t) = -\tilde{f}(k)T_2(t), \qquad \text{where} \qquad \tilde{f}(k) = f \int_{-D}^{+D} \cos(kx_1')w(|x_1'|)dx_1'. \qquad (4.9)$$

Thus, the response to perturbation is the same problem as we have addressed in Section 3.4, but with the classical Coulomb friction coefficient replaced by $\tilde{f}(k)$. Now, assuming that $w(|x|)$ is a function of bounded variation, we may assert on the basis of Fourier analysis that $|k\tilde{f}(k)| < B$, where B is some bound valid for all k. Thus $\tilde{f}(k) \to 0$ as $|k| \to \infty$.

Consider first cases for which the generalized Rayleigh wave does not exist. Then we showed that there is stability for sufficiently small f (Figs. 3.5 to 3.7). As $|k|$ increases, $\tilde{f}(k)$ will fall into such a range. Hence the weighted friction law gives stable response at sufficiently large $|k|$ and regularizes such cases. When the generalized Rayleigh wave does exist, the sliding problem is unstable to perturbation for all non-zero f, but we showed (eqs. (3.33) and (3.34)) that when f is changed slightly from zero, the real part of s is proportional to f, and hence the growth (or decay) rate is of order $\pm kf$. But the corresponding quantity $\pm k\tilde{f}(k)$ is bounded as $|k| \to \infty$ for the weighted friction law. This implies that the growth rate is bounded, so that we retain instability as $|k| \to \infty$ but secure well-posedness.

4.3 Regularization with a rate and state dependent friction law

Here, we study dynamic stability to linearized perturbations from a state of steady sliding along an interface between dissimilar materials with a rate and state dependent friction law at the interface. We show that the perturbations with high wavenumbers are always stable for sufficiently low sliding velocities when the instantaneous dependence of friction on sliding velocity is positive, thus giving rise to a well-posed stability problem.

We use the linear constitutive law (4.1) incorporating instantaneous and memory effects of both sliding velocity and normal stress on friction. As mentioned earlier, experimental evidence suggests a logarithmic dependence on sliding speed and a positive instantaneous dependence on the sliding speed at low speeds of the order of a few μm/sec. Using (4.1) in (3.22) with the constraint $\hat{D}_2 = 0$, we get an equation determining the slip perturbation in the form

$$\hat{D}_1 = \frac{(sq+1)\hat{Q}}{q\,|k|\,sg(s)}, \tag{4.10}$$

where, $q = |k|\,L/V_o$ as before and $g(s)$ is a function with a complicated form. Here, we are concerned only with the short wavelength behavior, $|k| \to \infty$, of the slip response. In that limit $g(s)$ takes the form

$$g(s) = (a\sigma_o/\mu V_o)s - Y_{11}(s) - (f - \alpha)Y_{21}(s). \tag{4.11}$$

where the dimensionless functions $Y_{ij}(s)$ are derived from the $\hat{M}_{ij}(s)$ of (3.20) by the relations

$$\hat{M}_{ij} = \mu\,|k|\,Y_{ij}(s). \tag{4.12}$$

and μ is a representative shear modulus.

First, consider the case where $a = 0$. Then $g(s)$ becomes $-Y_{11}(s) - (f - \alpha)Y_{21}(s)$ and is precisely of the form (3.28) derived in the Chapter 3 that governs stability to perturbations at a sliding dissimilar material interface with a Coulomb friction law. It follows from the analysis there that $g(s)$ has a zero with positive real part for a wide range of material combinations and values of f and α, giving rise to ill-posedness. In particular, the stability results depend on the existence of an interfacial wave, called the generalized Rayleigh wave (Weertman (1963), Achenbach and Epstein (1967), Gol'dshtein (1967)), in frictionless sliding of the two half-spaces. If the material parameters are such that the generalized Rayleigh wave exists in frictionless sliding, $g(s)$ has a zero in the right-half s-plane for any f and α as long as $(f - \alpha) \neq 0$. If the generalized Rayleigh wave does not exist for the bimaterial pair, a right-half plane zero exists if $|f - \alpha|$ is greater than a critical value, dependent on the bimaterial pair. The stability problem is therefore quite often ill-posed when $a = 0$.

Next, we let $a \neq 0$ and rewrite $g(s)$ in the form

$$g(P) = P/\epsilon + Y(P) \tag{4.13}$$

where $P = s/c_s$, $\epsilon = \mu V_o / a \sigma_o c_s > 0$, c_s is a representative shear wave speed and $Y(P) = -Y_{11}(P) - (f - \alpha)Y_{21}(P)$. In the following, we assume the sliding velocity V_o is sufficiently low that $\epsilon \ll 1$ Then, the roots of $g(P)$ must lie either close to the origin or close to a singularity of $Y(P)$. In the following, we show that all such possible roots have negative real parts.

It is easily seen that the root close to the origin will lie at $P = -\epsilon Y(0)$. Since $Y_{21}(0)$ is purely imaginary, the real part of the root is $-\epsilon Y_{11}(0)$. Observing that $Y_{11}(0)$ relates the static slip distribution in a single $\exp(ikx_1)$ mode to the shear stress in that mode at the interface, the requirement of a positive definite strain energy density function necessitates $Y_{11}(0) < 0$. Therefore, since $\epsilon > 0$, the root close to the origin always lies in the left-half P-plane.

Let us consider now the roots close to singularities of $Y(P)$. From the expressions (3.20), it is seen that $Y(P)$ is $O(P)$ as $P \to \infty$ and hence singular there (indeed, this is the singularity that gives rise to the radiation damping effect). Inspection of the form of $g(P)$ immediately informs that this singularity is not strong enough for a root to exist close to $P = \infty$. The only other condition when a singularity of $Y(P)$ can exist, if at all, would be for purely imaginary P (otherwise energy conservation would be violated for an interface with continuous displacements, hence no source of dissipation). That occurs when a Stonely wave exists for the bimaterial pair. That is a wave for which displacements are fully continuous, $\hat{D}_1 = \hat{D}_2 = 0$. From (3.17), it follows that the condition for the Stonely wave to exist is that the matrix $[\hat{K}_{ij}]$ be singular, i.e., that $D = \hat{K}_{11}\hat{K}_{22} - \hat{K}_{12}\hat{K}_{21} = 0$. When the Stonely wave exists, all the Y_{ij} are singular at $P = \pm i c_{St}/c_s$, where c_{St} is the speed of propagation of the Stonely wave.

We show in the remainder of this section that, in such cases, the root close to $P = \pm i c_{St}/c_s$ lies in the left-half P-plane if $\epsilon > 0$. Therefore, if the direct velocity effect component, a, of the friction law is positive and the sliding velocity is low enough that $\mu V_o / a \sigma_o c_s \ll 1$, the problem of stability to perturbations from steady sliding at a dissimilar material interface is always well-posed.

For material pairs for which c_{St} exists, one has the singular structure

$$Y_{11}(P) \sim \frac{iA}{i(c_{St}/c_s) - P}, \quad Y_{21}(P) \sim \frac{B}{i(c_{St}/c_s) - P} \tag{4.14}$$

near $P = \pm i c_{St}/c_s$, where A and B are real from expressions (3.18) and (3.20). Hence, when $\epsilon \ll 1$, there is a solution of $g(P) = 0$ near the pole given to leading order by

$$\frac{i(c_{St}/c_s)}{\epsilon} - \frac{iA}{i(c_{St}/c_s) - P} - \frac{(f - \alpha)B}{i(c_{St}/c_s) - P} \sim 0, \tag{4.15}$$

giving $P \sim i(c_{St}/c_s) - \epsilon(c_s/c_{St})[A - i(f - \alpha)B]$. Hence, provided that we can show that $A > 0$, the root lies in the left-half P-plane.

We now note, following Weertman (1963), Achenbach and Epstein (1967), and Gol'dshtein (1967), that relevant wave speeds for bimaterial problems are the generalized Rayleigh speed c_{GR}, for frictionless slip without opening, a companion wave speed c_{Op} for unimpeded opening without slip, and c_{St} for the case of there being neither slip nor opening. These speeds are taken as positive here and provide the roots $s = \pm i c_{GR}$, $s = \pm i c_{Op}$ and $s = \pm i c_{St}$ to the respective equations $\hat{K}_{22}(s) = 0$, $\hat{K}_{11}(s) = 0$ and $D(s) = 0$, when those equations have roots. Without loss of generality, we order the materials of Fig. 3.1 so that $c_{R1} < c_{R2}$. Then all these speeds c satisfy $c_{R1} < c < \min(c_{R2}, c_{s1})$. We note that along the imaginary axis $s = ic$, where c is real and $0 < c < \min(c_{s1}, c_{s2})$, $\hat{K}_{11}$ and $\hat{K}_{22}$ are real, whereas $\hat{K}_{21}$ is pure imaginary, which means that $D = \hat{K}_{11}\hat{K}_{22} - \left|\hat{K}_{12}\right|^2$ there. We also prove an ordering of the speeds as follows: Standing vibrations may be composed by superposing solutions of the type $\exp[ik(x_1 - ct)]$ and $\exp[ik(x_1 + ct)]$, and these have frequency $|k| c$. By Rayleigh's quotient, since the displacement field of the Stonely mode is kinematically admissible for the other two modes, it is of higher frequency and hence $\max(c_{GR}, c_{Op}) < c_{St}$. Study of the expressions (3.18) for the $\hat{K}_{ij}$, and the results of Weertman, Achenbach and Epstein, and Gol'dshtein mentioned above, then shows the following: (i) c_{Op} always exists, no matter what the bimaterial pair. (ii) c_{GR} exists only for bimaterial pairs that are not too dissimilar; it always exists if $c_{R2} < c_{s1}$ but goes out of existence if c_{R2} is too much larger than c_{s1}. (iii) c_{St} may exist only for bimaterial pairs for which c_{GR} exists. That follows because, if c_{GR} does not exist, which is the case when c_{R2} is sufficiently greater than c_{s1}, then examination of the expressions for the $\hat{K}_{ij}$ shows that $\hat{K}_{11} < 0$ and $\hat{K}_{22} > 0$ in the range $c_{Op} < c < c_{s1}$ in which c_{St} would have to lie, if it exists. But $\hat{K}_{11} < 0$ and $\hat{K}_{22} > 0$ imply that $D < 0$, so that $D = 0$ has no solution in that range, and hence c_{St} cannot exist.

Now, assuming that the bimaterial pair is such that c_{St} exists, which we have just seen to require that c_{GR} does also, the Stonely pole factor A will be positive if $\hat{M}_{11} = \hat{K}_{22}/D > 0$ for $\max(c_{GR}, c_{Op}) < c < c_{St}$. From the expressions for $\hat{K}_{11}$ and $\hat{K}_{22}$, and recalling that c_{Op} and c_{GR} are the respective zeros of $\hat{K}_{11}$ and $\hat{K}_{22}$, we have that $\hat{K}_{11}$ and $\hat{K}_{22}$ are of opposite sign for $\min(c_{GR}, c_{Op}) < c < \max(c_{GR}, c_{Op})$ and hence that $D < 0$ for c in that range. However, since c_{St} is the root of $D = 0$, that implies that $D < 0$ for $\max(c_{GR}, c_{Op}) < c < c_{St}$. In that same range of c, which is the range where c is greater than the roots of $\hat{K}_{11}$ and $\hat{K}_{22}$, both $\hat{K}_{11}$ and $\hat{K}_{22}$ are negative. Thus $\hat{K}_{22}/D > 0$ for $\max(c_{GR}, c_{Op}) < c < c_{St}$, which proves that $A > 0$ and hence that the root near the Stonely pole lies in the domain $\mathrm{Re}(P) < 0$. That shows, finally, that the problem of stability to perturbations from steady sliding at a dissimilar material interface is always well-posed at sliding velocity that is low enough that $\epsilon = \mu V_o/a\sigma_o c_s \ll 1$. This condition can, of course, be met only if the direct effect, a, of rate and state friction is present in the friction model.

The critical value of ϵ, ϵ_{cr} below which this result holds can be determined by finding value of ϵ_{cr} at which a root of $\epsilon = P/(Y_{11}(P) + (f - \alpha)Y_{21}(P))$ first enters the right half P-plane. If the material pair is such that the generalized Rayleigh wave exists in frictionless contact (i.e., $Y_{11}(P) = 0$) has a root at $P = \pm i c_{GR}/c_s$), this value is

$$\epsilon_{cr} = i(c_{GR}/c_s)/(f - \alpha)Y_{21}(P = i c_{GR}/c_s). \qquad (4.16)$$

Stability diagrams showing the value of ϵ_{cr} as a function of $(f - \alpha)$ for some bimaterial systems is shown in Figs. 4.4 and 4.5. Fig. 4.4 corresponds to a case where there is modest 20% mismatch in the densities of the solids while in Fig. 4.5 the density ratio is 5.0, typical of metal/polymer composite interfaces.

4.4 Estimates of a at high sliding speeds

As mentioned earlier, experimental studies of the velocity dependence of friction have mostly been at low sliding speeds of up to a few mm/sec. The logarithmic dependence of friction at these sliding speeds is well established and the parameter a is of the order of 0.01 for rocks in this speed range. In this section, we obtain estimates of the value of a at high sliding speeds for metals and rocks. Based on an experiment of Frutschy and Clifton (1997) involving sliding of a cutting tool material against a hard steel, a is shown to be about 0.2 at speeds of the order of a few m/sec. We also reinterpret prior results by Okubo and Dieterich (1986) on sliding of granite blocks at speeds of the order of a few cm/sec and estimate the velocity dependence of a in that case.

In the plate impact experiments of Frutschy and Clifton, normal and shear stress at a sliding interface between a tungsten carbide plate and a 4340 steel plate are altered rapidly (on the scale of a few nanoseconds) by wave reflections. A steady sliding state

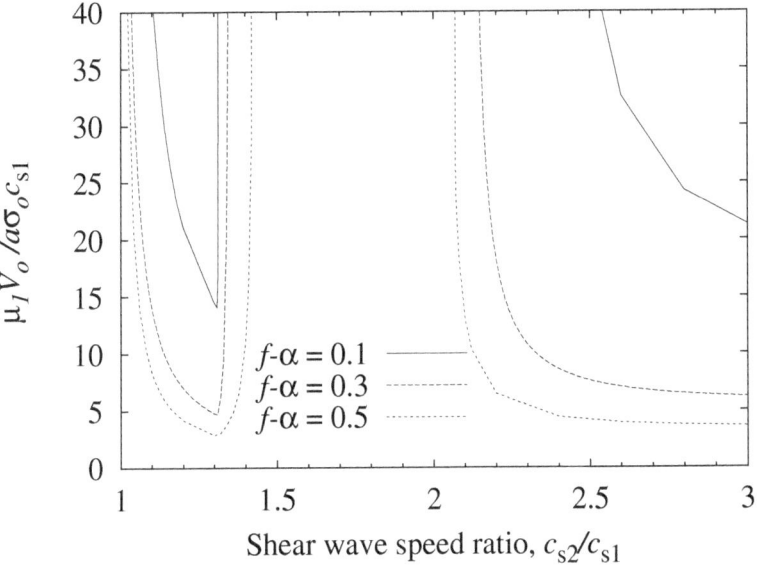

Figure 4.4: Critical value of $\epsilon = \mu V_o / a \sigma_o c_s$ for well-posedness with rate dependent friction for bimaterial pairs with $\rho_2/\rho_1 = 1.2$, $\nu_1 = \nu_2 = 0.25$

57

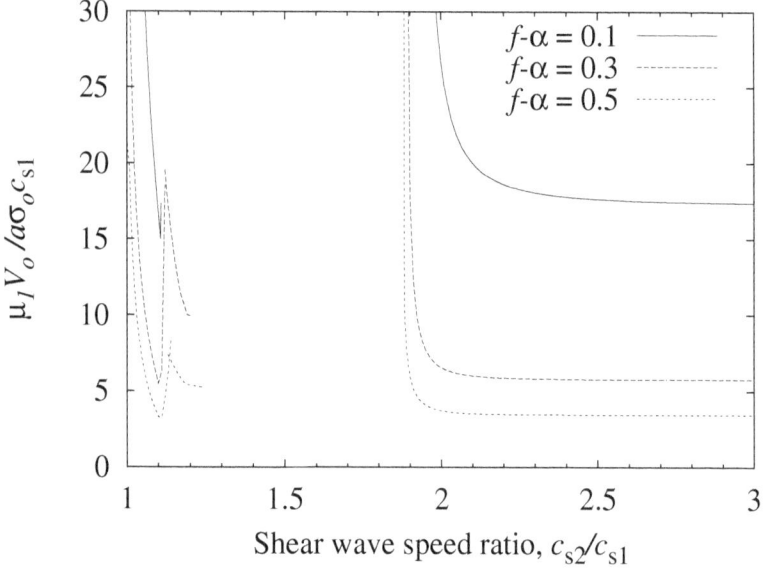

Figure 4.5: Critical value of $\epsilon = \mu V_o / a\sigma_o c_s$ for well-posedness with rate dependent friction for bimaterial pairs with $\rho_2/\rho_1 = 5.0$, $\nu_1 = 0.25$, $\nu_2 = 0.35$

with shear stress τ_1, compressive normal stress σ_1 and slip velocity V_1 is initialy setup by oblique impact of the steel flyer on the tungsten carbide target. The dilatational wave carrying the normal stress σ_1 is reflected onto the interface by the free surface of the target plate and the interface evolves to attain a new steady state with shear stress τ_2, normal stress σ_2 and slip velocity V_2. Subsequently, the shear wave initiated in the target plate during incidence of the flyer is reflected onto the interface by the free surface. This causes an instantaneous change in the sliding velocity and shear stress. From an experimental record of the interfacial response in shear wave loading, we attempt to estimate the friction parameters at these sliding speeds.

The shear wave carries a shear stress perturbation of $-\tau_1$ on reflection at the free surface. From one-dimensional wave theory, the shear stress τ and sliding velocity V at the interface when the reflected shear wave reaches the interface can be shown to be

$$\tau - \tau_2 = -C_t\tau_1 + (\mu/2c)(V_2 - V) \tag{4.17}$$

where C_t is the transmission coefficient of the interface and $\mu/2c$ is the radiation damping factor. If μ_f and μ_t denote shear moduli of the flyer and target plates, respectively and c_f and c_f are their respective shear wave speeds,

$$
\begin{aligned}
C_t &= \frac{2(\mu_f/c_f)}{(\mu_f/c_f) + (\mu_t/c_t)} \\
\mu/2c &= \frac{(\mu_f/c_f)(\mu_t/c_t)}{(\mu_f/c_f) + (\mu_t/c_t)}
\end{aligned} \tag{4.18}
$$

Further, we assume a friction law with a logarithmic dependence in velocity V and a state variable θ of the form introduced in Chapter 2 (Eqns 2.2 and 2.4) is operative at the interface

$$
\begin{aligned}
\tau - \tau_2 &= a\sigma_2\ln(V/V_2) + b\sigma_2\ln(V_2\theta/L) \\
\dot{\theta} &= -(V\theta/L)\ln(V\theta/L)
\end{aligned} \tag{4.19}
$$

Combining (4.17) and (4.19), and eliminating the state variable θ, the equation governing the evolution of the slip velocity V can be shown to be

$$[(a\sigma_2/V) + (\mu/2c)]\dot{V} = -(V/L)[(b - a)\sigma_2\ln(V/V_2) - C_t\tau_1 + (\mu/2c)(V_2 - V)] \tag{4.20}$$

The quantity $(b - a)$ can be determined from the steady state response of the interface. If τ_3 denotes the shear stress and V_3 the sliding speed at steady state, according to (4.19),

$$\tau_3 - \tau_2 = (b - a)\sigma_2\ln(V_2/V_3). \tag{4.21}$$

In the experiments, $\tau_1 = 0.21$ GPa, $\tau_2 = 0.10$ GPa, $\tau_3 = 0.10$ GPa, $V_2 = 14$ m/sec, $V_3 = 7$ m/sec and $\sigma_2 = 640$ MPa. This gives $(b - a) = 0.040$. The dependence on L in (4.20) can be eliminated by rescaling time. The properties of the flyer and target plates are taken to be $\mu_f = 78$ GPa, $\rho_f = 7800$ kg/m^3, $\mu_t = 262$ GPa and $\rho_t = 15000$ kg/m^3. This gives a transmission coefficient $C_t = 0.57$ and radiation damping factor $\mu/2c = 17.7$ MPa/(m/sec). The evolution of V and τ for various values of a is shown in Fig. 4.6. Comparing with the experimental data of Frutschy and Clifton (1997), we estimate a to be about 0.2.

Finally, we make estimates of a for granite at sliding speeds of up to a few cm/sec based on the work of Okubo and Dieterich (1986). They argue that while the logarithmic dependence on sliding rate is valid at low speeds, rate dependence of friction vanishes at high sliding speeds. They introduced a rate and state dependent friction law in the form

$$f = c_1/f_1 + (c_2/f_1)\log_{10}(c_3\theta + 1) - (c_1 f_2/f_1^2)\log_{10}(f_3/V + 1) \qquad (4.22)$$

where θ is a state variable and the constants c_1, f_2 and f_3 were obtained by fitting the friction law to experimental data. At low speeds, the term f_3/V dominates over the constant term in the third term above and the rate dependence is of the same form as in (4.19). However, at speeds much higher than f_3, the constant term dominates and there is effectively no rate dependence. Thus, this law has a high speed cutoff in the rate dependence. Linearizing the above friction law in the sliding speed about a state of steady sliding with velocity V_o and friction coefficient f_o, we get

$$f - f_o \sim \frac{c_1 f_2}{f_1^2} \frac{f_3}{V_o(V_o + f_3)}(V - V_o). \qquad (4.23)$$

For the friction law written in the form (4.19), the analogous term is $a(V - V_o)/V_o$ and hence we can identify

$$a = \frac{c_1 f_2}{f_1^2} \frac{f_3}{V_o + f_3}. \qquad (4.24)$$

Using the values of the constants c_1, f_1, f_2 and f_3 suggested by Okubo and Dieterich (1986), we get $a = 0.056/(1 + V_o/200)$ where V_o is in units of μm/sec.

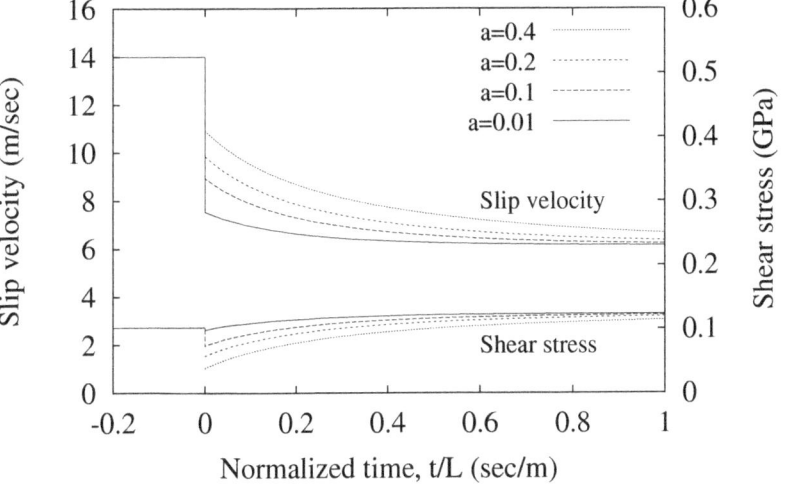

Figure 4.6: Simulation of shear stress and slip velocity evolution in the experiment of Frutschy and Clifton (1997)

References

1. Achenbach, J.D., 1973. Wave propagation in elastic solids. North-Holland, Amsterdam.

2. Achenbach, J.D., Epstein, H.I., 1967. Dynamic interaction of a layer and a half-space. Journal of the Engineering Mechanics Division EM5, 27-42.

3. Adams, G.G., 1995. Self-excited oscillations of two elastic half-spaces sliding with a constant coefficient of friction. J. Appl. Mech. 62, 867-872.

4. Adams, G.G., 1998. Steady sliding of two elastic half-spaces with friction reduction due to interface stick-slip. J. Appl. Mech. 65, 470-475.

5. Adams, G.G., 2000. Radiation of body waves induced by the sliding of an elastic half-space against a rigid surface. J. Appl. Mech. 67, 1-5.

6. Andrews, D.J., Ben-Zion, Y., 1997. Wrinkle-like slip pulse on a fault between different materials. J. Geophys. Res. 102, 553-571.

7. Beeler, N.M., Tullis, T.E., Weeks, J.D., 1994. The roles of time and displacement in the evolution effect in rock friction. Geophys. Res. Lett. 21, 1987-1990.

8. Ben-Zion, Y., Andrews, D.J., 1998. Properties and implications of dynamic rupture along a material interface. Bull. Seismol. Soc. Amer. 88, 1085-1094.

9. Breitenfeld, M.S., Geubelle, P.H., 1998. Numerical analysis of dynamic debonding under 2D in-plane and 3D loading. Int. J. Fracture 93, 13-38.

10. Cochard, A., Rice, J.R., 2000. Fault rupture between dissimilar materials: Ill-posedness, regularization, and slip-pulse response. J. Geophys. Res. 105, 25891-25907.

11. Dieterich, J.H., 1978. Time-dependent friction and the mechanism of stick-slip. Pure Appl. Geophys. 117, 190-806.

12. Dieterich, J.H., 1979. Modeling of rock friction - 1. Experimental results and constitutive equations. J. Geophys. Res. 84, 2161-2168.

13. Dieterich, J.H., 1981. Constitutive properties of faults with simulated gouge. In: Carter, N. L., Friedman, M., Logan, J. M., Stearns, D. W. (Eds.), Mechanical Behavior of Crustal Rocks, Geophys. Monogr. Ser. 24, 103-120. Amer. Geophys. Un., Washington, DC.

14. Dieterich, J.H., 1994. A constitutive law for rate of earthquake production and its application to earthquake clustering. J. Geophys. Res. 99, 2601-2618.

15. Frutschy, K.J., Clifton, R.J., 1997. Plate-impact technique for measuring dynamic friction at high temperatures. J. Tribology 119, 590-593.

16. Geubelle, P.H., Rice, J.R., 1995. A spectral method for three-dimensional elastodynamic fracture problems. J. Mech. Phys. Solids 43, 1791-1824.

17. Gol'dshtein, R.V., 1967. On surface waves in joined elastic media and their relation to crack propagation along the junction. PMM 31(3), 468-475. (English translation, Appl. Math. Mech. 31, 496-502, 1967).

18. Gu, J.-C., Rice, J.R., Ruina, A.L., Tse, S.T., 1984. Slip motion and stability of a single degree of freedom elastic system with rate and state dependent friction. J. Mech. Phys. Solids 32, 167-196.

19. Harris, R.A., Day, S.M., 1997. Effects of a low-velocity zone on a dynamic rupture. Bull. Seismol. Soc. Amer. 87, 1267-1280.

20. Lapusta, N., 2001. Elastodynamic analysis of sliding with rate and state friction. Ph.D. Thesis, Division of Engineering and Applied Sciences, Harvard University.

21. Linker, M.F., Dieterich, J.H., 1992. Effects of variable normal stress on rock friction: Observations and constitutive equations. J. Geophys. Res. 97, 4923-4940.

22. Marone, C., 1998. Laboratory-derived friction laws and their application to seismic faulting. Ann. Rev. Earth Planetary Sci. 26, 643-696.

23. Martins, J.A.C., Guimaraes, J., Faria, L.O., 1995. Dynamic surface solutions in linear elasticity and viscoelasticity with frictional boundary conditions. Journal of Vibration and Acoustics, Trans. ASME 117, 445-451.

24. Martins, J.A.C., Simões, F.M.F., 1995. On some sources of instability/ill-posedness in elasticity problems with Coulomb friction. In: Raous, M., Jean, M., Moreau, J.J. (Eds.), Contact Mechanics. Plenum Press, New York, pp. 95-106.

25. Okubo, P.G, Dieterich, J.H., 1986. State variable fault constitutive relations for dynamic slip. In: Das, S., Boatwright, J. and Scholz, C.H. (Eds.),Earthquake Source Mechanics, Geophys. Monogr. Ser., Vol. 37., pp. 25-35.

26. Perrin, G., Rice, J.R., Zheng, G., 1995. Self-healing slip pulse on a frictional surface. J. Mech. Phys. Solids 43, 1461-1495.

27. Prakash, V., Clifton, R.J., 1993. Time resolved dynamic friction measurements in pressure-shear. In: Ramesh, K.T. (Ed.), Experimental Techniques in the Dynamics of Deformable Solids, ASME, AMD-Vol.165, New York, pp. 33-48.

28. Prakash, V., 1998. Frictional response of sliding interfaces subjected to time varying normal pressures. Journal of Tribology, Trans. ASME 120, 97-102.

29. Renardy, M., 1992. Ill-posedness at the boundary for elastic solids sliding under Coulomb friction. Journal of Elasticity 27, 281-287.

30. Rice, J.R., Ben-Zion, Y., 1996. Slip complexity in earthquake fault models. Proc. Natl. Acad. Sci. USA 93, 3811-3818.

31. Rice, J.R., Gu, J.-C., 1983. Earthquake aftereffects and triggered seismic phenomena. Pure Appl. Geophys. 121, 187-219.

32. Rice, J.R., Ruina, A.L., 1983. Stability of steady frictional slipping. Trans. ASME J. Appl. Mech. 50, 343-349.

33. Richardson, E., Marone, C., 1999. Effects of normal force vibrations on frictional healing. J. Geophys. Res. 104, 28859-28878.

34. Ruina, A.L., 1983. Slip instability and state variable friction laws. J.Geophys Res. 88, 10359-10370.

35. Simões, F.M.F., Martins, J.A.C., 1998. Instability and ill-posedness in some friction problems. Int. J. Eng. Sci. 36, 1265-1293.

36. Tullis, T.E., Weeks, J.D., 1986. Constitutive behavior and stability of frictional sliding in granite. Pure Appl. Geophys. 124, 383-414.

37. Weertman, J., 1963. Dislocations moving uniformly on the interface between isotropic media of different elastic properties. J. Mech. Phys. Solids 11, 197-204.

38. Weertman, J., 1980. Unstable slippage across a fault that separates elastic media of different elastic constants. J. Geophys. Res. 85, 1455-1461.